Praise for *An Optimist's Tour of the Future*

"From longevity science to robotics to cancer research, Stevenson explores the most cutting-edge ideas in science and technology from around the world, the important ethical and philosophical questions they raise, and, perhaps most importantly, the incredible potential for innovation through the cross-pollination of these different ideas and disciplines." —*THE ATLANTIC*

"Stevenson does a great job of delivering all the facts with a sense of wonder and curiosity. . . . I highly recommend this book." —*WIRED'S* GEEK DAD

"Part trend-hunting, part rigorous research, part cultural anthropology, *An Optimist's Tour of the Future* may just be our generation's version of Bill Bryson's iconic *A Short History of Nearly Everything*—a bold and entertaining blueprint for a future that's ours to shape and ours to live." —BRAINPICKINGS.ORG

"A series of sharp and fascinating interviews with technological innovators and scientific visionaries." —*THE WALL STREET JOURNAL*

"Stevenson surveys a huge number of research fields—synthetic biology, nanotech, robotics and alternative energy to name just a few—in labs on nearly every continent, and weaves the most promising aspects of each into one grand vision. . . . The book is a refreshing reminder that the future will always belong to the optimists." —*THE NEW SCIENTIST*

"Stevenson describes our future's possibilities with a journalist's eye for detail, a teacher's knack for translating complexities, and a comic's wry commentary." —*THE CHRISTIAN SCIENCE MONITOR*

"From locations as diverse as the Maldives, the Australian Outback and a hotel in space, to subjects as diverse as nanotechnology, synthetic biology and philosophy, there's seemingly nothing off limits in Mark Stevenson's ambitious new book." —CBSNEWS.COM

"Stellar reviews all over the place, and it is certainly one of the most interesting science books we've read for a long time . . . contentious subjects tackled in an approachable way." —*NEW STATESMAN*

". . . a rollicking roller-coaster ride around the cutting edge of science with dozens of laugh-out-loud moments. His access to some of the planet's brainiest scientists might suggest that he himself has a professional reputation of some stature. But having cornered them in the labs, Stevenson is far from star-struck." —*THE SCOTSMAN*

"Out there on the edge of research, Stevenson puts the trends of tomorrow's world into perspective with a quizzical, fast-paced, quick-witted tour of the scientific horizon." —*THE TIMES* (LONDON)

"[A] grand tour of charismatic technologies and their prophets . . . Stevenson bags an impressive list of interviewees." —*FINANCIAL TIMES*

"Infectiously enthusiastic . . . Despite the book's title Stevenson is not a relentlessly wide-eyed optimist. He is alive to the self-puffing of the scientists and engineers he meets, and has a nice eye for the absurd . . . plenty of material for pub speculation." —*SUNDAY TIMES*

"Stevenson's journey is a measured effort to take stock of the reasons for hope, and to keep faith with the enlightenment project. You don't have to find all the people he meets persuasive, or go along with all his judgments of their projects. But he makes a good case for believing that we can have a future worth making an effort to reach." —*THE GUARDIAN*

"Insightful . . . entertaining . . . Stevenson's restrained and rational enthusiasm makes for optimistic reading." —*THE AUSTRALIAN*

"Blind optimism is the last thing the world needs—it's a recipe for disappointment. But what about optimism based on careful reasoning? That's what's on offer here. Stevenson's tools are curiosity, open-mindedness, clarity, and reason. That makes his journey intriguing . . . and ultimately exhilarating." —CHRIS ANDERSON, TED CURATOR

"Mark Stevenson is a futurist endowed with abundant optimism. Where some fear that robotics and artificial intelligence will dehumanize the world, he sees an enhancement of freedom; where some see nanotechnology as an army of sub-micron Lilliputians taking over, he sees an invisible array of artisans performing miracles in tight spaces. It's invigorating to be led to the far-side by someone who sees the bright-side!" —HENRY POLLACK, AUTHOR OF *A WORLD WITHOUT ICE*

"This is a brilliant book, and Mark Stevenson is the perfect guide to a dizzying future that is already here. Genetic innovation. Social robots. Nanofactories. The ideas come so quickly, with such great humor—it's like the smartest dinner party you've ever attended." —PETER MILLER, AUTHOR OF *THE SMART SWARM*

"Uplifting and liberating. Mark Stevenson is the perfect tour guide: optimistic without being naive, fun but not silly, entertaining yet also an educator. By the end I was utterly convinced that, given the opportunity to flourish, human ingenuity can disperse the clouds that hang over us today. This is a book to gladden the gloomiest heart." —MICHAEL BROOKS, AUTHOR OF *13 THINGS THAT DON'T MAKE SENSE*

"Might also be called 'a tour of the future by someone with great insight, quick wit, and the ability to connect the dots—even the vague, moving dots of the future.'" —KUSF-AM (SAN FRANCISCO)

AN
OPTIMIST'S TOUR
OF THE
FUTURE

One Curious Man Sets Out to Answer "What's Next?"

MARK STEVENSON

AVERY   A MEMBER OF PENGUIN GROUP (USA) INC.   NEW YORK

Published by the Penguin Group
Penguin Group (USA) Inc., 375 Hudson Street, New York, New York 10014, USA • Penguin
Group (Canada), 90 Eglinton Avenue East, Suite 700, Toronto, Ontario M4P 2Y3, Canada
(a division of Pearson Penguin Canada Inc.) • Penguin Books Ltd, 80 Strand, London
WC2R 0RL, England • Penguin Ireland, 25 St Stephen's Green, Dublin 2, Ireland (a division
of Penguin Books Ltd) • Penguin Group (Australia), 250 Camberwell Road, Camberwell,
Victoria 3124, Australia (a division of Pearson Australia Group Pty Ltd) • Penguin Books
India Pvt Ltd, 11 Community Centre, Panchsheel Park, New Delhi–110 017, India •
Penguin Group (NZ), 67 Apollo Drive, Rosedale, North Shore 0632, New Zealand
(a division of Pearson New Zealand Ltd) • Penguin Books (South Africa) (Pty) Ltd,
24 Sturdee Avenue, Rosebank, Johannesburg 2196, South Africa

Penguin Books Ltd, Registered Offices: 80 Strand, London WC2R 0RL, England

First trade paperback edition 2012
Copyright © 2011 by Mark Stevenson

Most Avery books are available at special quantity discounts for bulk purchase for sales
promotions, premiums, fund-raising, and educational needs. Special books or book excerpts
also can be created to fit specific needs. For details, write Penguin Group (USA) Inc.
Special Markets, 375 Hudson Street, New York, NY 10014.

The Library of Congress catalogued the hardcover edition as follows:

Stevenson, Mark, date.
   An optimist's tour of the future : one curious man sets out to answer "what's next?" / Mark Stevenson.
      p.   cm.
   ISBN 978-1-58333-414-0
 1. Research—Forecasting.   2. Technological forecasting.   3. Technological
innovations—Forecasting.   4. Science—Humor.   I. Title.
   Q180.A1S74   2011   2010039040
   601'.12—dc22

ISBN 978-1-58333-456-0 (paperback edition)

Printed in the United States of America
10 9 8 7 6 5 4 3 2 1

BOOK DESIGN BY MEIGHAN CAVANAUGH

While the author has made every effort to provide accurate telephone numbers and Internet addresses
at the time of publication, neither the publisher nor the author assumes any responsibility for errors,
or for changes that occur after publication. Further, the publisher does not have any control over
and does not assume any responsibility for author or third-party websites or their content.

For Caroline

CONTENTS

Somewhere, something incredible is waiting to be known.

—*Carl Sagan*

The future is here. It's just not widely distributed yet.

—*William Gibson*

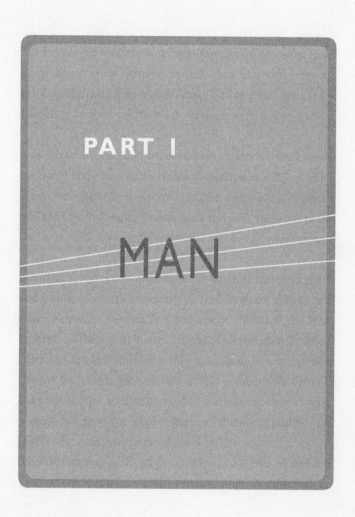

PART I

MAN

"The World's Most Dangerous Idea"

Getting older is no problem. You just have to live long enough.

—*Groucho Marx*

'm on a train to Oxford wondering how long I am going to live.

This isn't because I think we're about to crash, nor is it a reaction to the sandwich First Great Western has just sold me. It's because a few weeks ago the stalking horse of Mortality popped into my head and, without invitation, asked, "What are you going to do with the rest of your life, then?" (Mortality, it turns out, sounds a lot like my father.) Whatever the answer, one thing I do know is that my future will play out in a world different to the one we've all grown up in.

"The times they are a-changin'," sang Bob Dylan as the sixties got into the full flow of social revolution. Fifty years later there's another revolution going on: the population is rocketing, the planet is becoming urban (over half of us now live in cities), medicine is curing the previously incurable, ninety-year-olds take parachute jumps,[1] spaceships are owned by businessmen, the climate is shifting and the world's knowledge is becoming available to anyone with an Internet connection.

So I'm setting out on a journey that I hope will tell me what the world of my future will look like. Are we going to cure cancer? What is the "biotech revolution"? Are the robots finally coming? What's all that nanotechnology stuff about? How will the Internet continue to shape society?

If the climate is changing, how will that affect us and can we do anything about it? Are we going back to space? Is technology going to be our friend or run off into the distance, leaving us dazed and confused? And how will the answers to all these questions tangle together and change the way we live, work and play? I want to find an answer to the most personal and yet the biggest of questions, a question all of us ask:

"What's next?"

But first I need to work out how much "next" I'm likely to see. Or, to put it another way—how long have I got? How long is my future?

The UK Office for National Statistics tells me that average life expectancy at birth for a man born in Great Britain in 1971 (that's me) was just over sixty-nine years.[2] (If I'd been born in the U.S. the figure is sixty-eight.)[3] They also inform me that, as a thirty-nine-year-old male in 2010, I can now expect to live ten years longer.[4] (In the U.S. I would have gained another nine years).[5] The mere act of staying alive, it seems, has given me more living to do.

"Life expectancy creep" is well documented. Average age at death has been steadily edging upward by roughly a quarter of a year for every calendar year that's passed since vaguely reliable records began (around the mid-nineteenth century). Some say this trend is likely to continue (even accelerate) in tandem with medical advances, and that centenarians will soon become commonplace.[6] Others argue that the rise in statistical life expectancy is due more to rescuing the young (by fighting infant tuberculosis, for example). Hinting at a natural limit to our longevity, Stuart Jay Olshansky of the University of Illinois School of Public Health points out, "You can't save the young twice."[7]

With the notable exceptions of Ozzy Osbourne, Keith Richards and all of Aerosmith, there is also a well-established link between the way someone lives and their expected life span. Accordingly I've answered a variety of lifestyle and family history questions at the request of several

Internet-based life expectancy calculators. My favorite, "Death Clock," produces a personal "death date" based on "normal," "optimistic," "pessimistic" or "sadistic" calculations (the latter announcing you're already dead) before displaying the number of seconds you have left to live . . . and then *counting down*. There's something ironically mesmerizing about literally watching your life tick away. . . . I wonder if anyone's actually *died* doing it?

The consensus is that, if I maintain my current lifestyle, I'll live past eighty but not eighty-five (so, roughly in line with government statistics). However, if I improve my diet, get more exercise, continue working (according to some studies, a key strategy for keeping the grim reaper at bay) and drink less, I've a good chance of pushing past ninety. And if life expectancies keep creeping up the way they have been, fifty years from now I could have up to *another* twelve years on the planet. It's not infeasible I'll make it past a hundred. I find this encouraging: I'm still single with no kids, but it appears there's plenty of time to raise a family—and maybe even understand cricket.

My first port of call, Nick Bostrom, founder of Oxford University's Future of Humanity Institute, quite agrees. In fact, he thinks I could live, not for a hundred years, but for thousands. And he's not joking.

Bostrom is an advocate for transhumanism. Described as "the world's most dangerous idea,"[8] the concept got its modern name in *Religion Without Revelation*, written in 1927 by Julian Huxley, a leading biologist, the first director-general of UNESCO and founder of the World Wildlife Fund. Busy chap.

He wrote:

> The human species can, if it wishes, transcend itself—not just sporadically, an individual here in one way, an individual there in another way—but in its entirety, as humanity. We need a name for this new belief.[9]

The term Huxley came up with was "Transhumanism"—the idea that man could remain essentially human but transcend what nature had given him. Skip forward seventy-seven years to 2004 and Aubrey de Grey, the controversial biogerontologist, declares, "I think the first person to live to a thousand might be sixty already."[10] I'm traveling to see Nick because the debate around transhumanism is a car-crash of science, ethics and social policy. It asks us to rethink what medicine is, what society might look like and how far we should go with our technology. I'm kind of throwing myself in at the deep-end.

Via a stuttering wi-fi connection, I use part of my journey to look again at a typical polemic given by de Grey at the 2004 Technology, Entertainment, Design (TED) conference in California. In it he asks if anyone in his audience "is in favor of malaria." When no one raises their hand he continues, "I put it to you that the main reason we think malaria is a bad thing is because of a characteristic it shares with aging—and here is that characteristic"—at which point he shows a slide bearing the words "Because it kills people!!!!!!" "The only real difference is that aging kills considerably *more* people." De Grey proceeds to compare aging to fox hunting, claiming they are both "traditional," "keep the numbers down" and, his punch line, "fundamentally barbaric."[11] At the same conference, Nick Bostrom later stated, "Death is a big problem. If you look at the statistics the odds are not very favorable. So far most people who have lived have also died."[12]

But banishing aging is only one half of the transhumanists' truly radical agenda. They also advocate a new vision of the human being: the transhuman—a human not only saved from aging, but also *enhanced* beyond our current "biological limitations."

As a child I used to play with a transhuman called Steve Austin. Steve was a toy doll of the Six Million Dollar Man, a merchandising spin-off from the TV show of the same name. Just like TV Steve, my doll had "bionic" body parts that helped him see further, run faster and punch harder than his (frequently mustachioed) adversaries. His manufactured body parts were replacements for those lost in a plane crash. Over the opening credits his backstory was intoned: "We have the capability to make the world's first bionic man . . . Better, stronger, faster."

These last words reflect much of the transhumanists' enhancement credo—that transhumanism isn't only about keeping us alive and healthier for longer, but allowing us to *be more*. Be better. Be stronger. Be faster.

History tells us of numerous individuals who've searched for the secret of living forever. Chinese alchemist Ge Hong (born in 283 CE) believed immortality was achieved by cultivating everlasting union with yourself, allowing you to sip from a wellspring of a "metaphysical oneness" or *xuan,* which permeated all things.[13] He died.

Today, numerous purveyors of quackery peddle a baffling variety of what they claim are immortality-delivering products. San Francisco–based Alex Chui will sell you a pair of magnetic finger rings or "eternal life devices" for around thirty dollars that even come with a ninety-day refund if you're not satisfied. This strikes me as an extraordinarily unconfident guarantee for something that promises everlasting life. When I contact Chui to raise this point he informs me, without irony, that "a lifetime guarantee is illegal. You cannot make such a claim." Perhaps the best-known "scientific" method of trying to avoid the great gig in the sky is "cryonic suspension"—the process of preserving the recently deceased by freezing their body or, for those with less funds (and I find this deliciously macabre), just their head. This method rests on the belief that physicians in the future will be able to thaw out, reanimate and cure these "cryonauts" (or, in the parlance of some wittier cryonics enthusiasts, "the metabolically challenged") of whatever it was that killed them in the first place. It's fair to say that cryonics hasn't been accepted as a mainstream medical practice. As far as today's medicine is concerned, death is a one-way trip.

I mention these examples because stories of misguided optimism, wishful thinking, moral crusading, faith in unproven technology and, let's be honest here, out-and-out fruitcake-ism are not uncommon in the history of man's quest to cheat death. Accordingly, I'm approaching the transhumanist project with a good dose of skepticism. I'm not the only one to be dubious.

Fed up with what they perceived as a lot of media "bad science" stirred up by the anti-aging part of the transhumanist project, in 2006 the *MIT Technology Review* offered a $20,000 prize to anyone who could demonstrate that Aubrey de Grey's ideas and research were "unworthy of learned debate," $10,000 of which was supplied by de Grey himself. (Jason Pontin, the *Technology Review*'s editor, had previously written a stinging profile of de Grey, characterizing him as a shabby, beer-addicted troll and comparing his work to "science fiction."[14])

The submission that attracted the most praise (written by Preston Estep and colleagues) claimed that de Grey's work enjoys notoriety "due almost entirely to its emotional appeal" and dismissed much of it as "pseudoscience." Between scientists that's the equivalent of calling Muhammad Ali "a sissy."

The judges (who included world experts in artificial intelligence, nano-technology, pathology and genetics), however, didn't think Estep and his colleagues had done enough. While their feeling was that de Grey's ideas exist "in a kind of antechamber of science, where they wait (possibly in vain) for independent verification," they're also "not demonstrably wrong." Judge Nathan Myhrvold wrote that "it is a hallmark of the scientific process that it is fair about considering new propositions; every now and then, radical ideas turn out to be true. Indeed, these exceptions are often the most momentous discoveries in science."[15]

History is, of course, littered with mavericks initially ridiculed or dismissed and later (sometimes *much* later) accepted as geniuses, including Charles Darwin (evolution), Gregor Mendel (genetic inheritance), Robert Goddard (liquid-fueled rockets), Louis Pasteur (germ theory) and the Wright brothers (powered flight). Then again, the past is also full of challenges to accepted wisdom which crashed and burned. I give you James McConnell and Georges Ungar, who believed that memories were encoded in molecules and could therefore be transferred from one animal to another—giving rise to the possibility that you could take a pill and go on to recall the complete works of Shakespeare.[16] No, I'd never heard of them either.*

In summary, Bostrom and de Grey *might* be onto something. And if the transhumanists really are the outriders of the future, that, as a neuroscientist friend of mine observed, "is a bit of a headfuck."

Nick Bostrom's self-penned biography reads:

> Beside philosophy, I also have a background in physics, computational neuroscience, mathematical logic, and artificial intelligence. My performance as an undergraduate set a national record in Sweden. I was a busy young man. Before becoming a tweedy academic, I also dabbled in painting and poetry, and for a while I did stand-up comedy in London.[17]

He is thirty-six. I'd hate to read his biography when he's fifty, by which time he will no doubt have enjoyed a brief stint in the West End as a dancer, proved the Riemann hypothesis (one of the great unsolved problems in mathematics), built a robot wife and published several successful cookbooks.

Transhumanism raises deep philosophical and social questions. What will the impact on society and personal relationships be in a world where we might live forever? Would any of us get married if we knew it was for a thousand years? When do we retire? Isn't there a massive overpopulation problem? Won't we run out of natural resources? If we merge with our machines, what might we lose? What about pensions? What will the AARP do? Anyway, death is natural, isn't it?—part of the cycle by which our species evolves? Isn't aging a defining characteristic of being human? Wouldn't we go mad? And even if we could defeat death, should we?

With middle age at my back the idea of being rejuvenated to a younger

---

*To be fair, McConnell and Ungar were never conclusively proved wrong, but enthusiasm and funding for their research ran out when the results they did publish proved less than compelling.

body is very attractive. But I worry that what I think might be good for me might not be good for humankind as a whole. To their credit, most transhumanists accept that ending aging (if it's possible) won't arrive without massive upheavals to society. The question Aubrey de Grey asks is, Do the risks "outweigh the downside of doing the opposite, namely leaving aging as it is . . . condemning a hundred thousand people a day to an unnecessarily early death? You know, if you haven't got an argument that's that strong then just don't waste my time." For transhumanists this is a moral argument.

On the issue of overpopulation Aubrey de Grey has written, "Basically our options are extremely simple: either restrict the birth rate or raise the death rate," by which he means if we all start living longer and keep to one planet, we'll probably have to breed less. Bostrom is more circumspect. Such questions, he has stated, should "be addressed in a sober, disinterested way, using critical reason and our best available scientific evidence" in order to decide "what policies it makes sense for humanity to pursue." I admire his sentiments, but I wonder if asking us to think about all this rationally is like asking a fly to consider the aesthetic qualities of a windshield. Still, this is why I've come to Oxford.

In the same week I visit Professor Bostrom, two news stories catch my eye. The first concerns scientists at the Spanish National Cancer Center in Madrid who have been breeding mice which live, on average, *thirty percent longer* than their "normal" brethren.[18] Led by Maria Blasco, the team is building on the work of Elizabeth Blackburn, Carol Greider and Jack Szostak, who discovered a startling substance (strictly an enzyme*) called "telomerase" while they were investigating a micro-critter named "Tetrahymena" found in freshwater ponds, a discovery for which the three shared the 2009 Nobel Prize for Medicine.[19]

Tetrahymena has two particularly interesting characteristics. One, it has *seven* sexes, which I can only imagine means they swing. A lot. Two,

---

*A substance cells use to spur on chemical reactions—a "biological catalyst."

it's biologically immortal. You read that right. If it can eat, it just keeps on living—and telomerase is one of the reasons why.

Tetrahymena isn't subject to something called the "Hayflick limit," named after medical microbiologist Leonard Hayflick. Along with his colleague Dr. Paul Moorhead, Hayflick worked out that human cells have a built-in "age-then-die" program that kicks in after they've divided somewhere between forty and sixty times. When the cell reaches this limit, it enters its last days—a state called "senescence." (De Grey calls his assault on aging "Strategies for Engineered Negligible Senescence.") But a constant supply of telomerase (which Tetrahymena continually replenish themselves with) can keep resetting a cell's "divide count." This means that even if a cell is fifty generations down the line, it behaves like it's only divided once (keeping it well away from the death knell of the Hayflick limit).*

Human telomerase levels decline as we get older, so the natural question for the anti-aging brigade is, If we increase our supply of telomerase, can we combat aging? Several studies have shown that people who live healthy lifestyles (and therefore tend to live longer) have higher levels of telomerase in certain cell lines, supporting a direct link between longevity and telomerase—and increasing the supply of it to Blasco's mice did indeed dramatically lengthen their lives.[20] Against the constant backdrop of continuing medical advances, stories like this seem to suggest the transhumanists may not be quite as crazy as they might first appear.

But there's always a catch. I subsequently discover that cancer cells also come with a helpful supply of telomerase—which is one of the reasons they divide and multiply with such awful success. The telomerase helps keep them from entering senescence too. Cancer, after all, is defined as the uncontrollable division of cells. (That's what tumors are—cancerous cells that have gone on a dividing frenzy without quitting.) So, if you go for a telomerase top-up, like Blasco's mice, you'd need to be bred cancer-resistant too.

---

*In deference to Professor Hayflick, it's only right to point out that he is a strong critic of de Grey's philosophy. He does not believe aging "is a disease" and thinks that an "anti-aging" stance is contradictory to being human.

The second story to grab my attention that week is even more compelling.

In 2008, Claudia Castillo suffered a collapse of part of the windpipe leading to her left lung. The result was a debilitating shortness of breath, making it impossible for her to look after her children, to work, clean, shop or cook.[21]

The usual medical options would be wholesale removal of a lung or a partial trachea (windpipe) transplant. Transplants are particularly tricky because our immune systems tend to reject donated body parts, requiring patients to take immunosuppressive drugs. These reduce your body's desire to reject the new part, but at the same time can severely hamper its ability to combat infection. Instead, aged thirty, Castillo became a pioneer for "stem cell" therapy.

Keeping up with advances in stem cell therapy is like trying to run after a Formula 1 car with your shoelaces tied together. The field is moving at an incredible rate. Stem cells are cells that are yet to pick a career—or to put it another way they are "blank" cells, which can ultimately be "programmed to perform particular tasks."[22] This is why human embryos naturally have a ready supply and is where researchers have often got them from (leading to all that controversy). But adults also carry around their own supply of "tissue-specific" stem cells. These chaps are also still waiting to be "programmed" but have a smaller palette of career choices, based on their tissue of origin. That makes them less controversial (no baby parts) but less widely applicable (they can only turn into particular bits of you). It's now possible, however, to take ordinary cells (say, from the skin on your hand) and "reboot" them back to their earlier stem cell selves, doing away with many ethical concerns altogether.[23] But even that may not be necessary. Early in 2010, scientists at Stanford University in California announced they had managed to convert mouse skin cells *directly* into mouse brain cells (with no intermediate stem cell phase).[24] A few months later a team from Harvard declared they could convert human blood cells to stem cells that may have the ability to grow into any kind of tissue.[25]

Castillo received a donated trachea that had been stripped of cells from the donor (leaving just a collagen pipe). This was then "seeded" with stem cells harvested from her own body (and nurtured by a team at the University of Bristol) that grew into tissue. The result? Her body accepted the new windpipe as its own and medical history was made.* It sounds like science fiction, but the patient is out dancing.

Castillo isn't the only person wandering around with a stem cell–grown body part. Kaitlyne McNamara's doctors isolated healthy adult stem cells from the diseased bladder she was born with and used them to grow an entire, fully functioning (and healthy) replacement, which they then implanted.[26] The Wake Forest Institute for Regenerative Medicine in Winston-Salem, North Carolina, "are working on more than 22 different organs and tissues"[27] including kidneys, livers, retinas and muscle. Doris Taylor, Harold Ott and colleagues at the Center for Cardiovascular Repair at the University of Minnesota took the collagen shell of a rat's heart, sprayed it with stem cells and it started beating.[28] "When we saw the first contractions we were speechless," says Ott. In February 2010 the technology was licensed to a company called Miromatrix Medical, which hopes to revolutionize organ transplants with the technique.[29]

There are three fronts on the transhumanists' war with our inbuilt human limitations. The first is an attack on the restrictions of our biology. If the transhumanists wanted a field marshal to lead this advance, then Aubrey de Grey would be the obvious candidate. He's encouraging the scientific community to find solutions to "seven major types of molecular and cellular damage that eventually become bad for us" which

---

*An interesting coda to Claudia's story is that she nearly didn't get her operation. With a time limit of sixteen hours for Claudia's lab-nurtured cells to reach their destination, the Bristol team booked the only direct Bristol-to-Barcelona flight—operated by easyJet. Despite getting agreement from easyJet several months in advance that student Philip Jungerbluth could carry Claudia's cells as hand luggage, check-in staff flatly refused to allow the material onto the plane, citing "anti-terror" security measures. Professor Martin Birchall was nearly arrested by armed police as he fought vainly against the orange-uniformed face of low-cost air travel. The answer? A friend of Philip's (a pilot) and €11,000 out of Birchall's own pocket to charter a private flight. Philip tells me easyJet wouldn't even refund the €70 for the original ticket. Insert your own chain of expletives here to express how outrageous that is.

he believes, once fixed, will put an end to aging. He's a big fan of stem cells and long-lived mice (his Methuselah Foundation offers prizes for scientific research teams who extend mouse life spans). De Grey's line of attack could be called the "wet front"—dealing with the slippery world of cells and their biology.

The second front could be called the "dry" front—evolving as it does from our experiences building machines. Here the transhumanists could reasonably propose Ray Kurzweil (a man I add to my list of those I should meet later on my journey) as the commander in chief—one of the world's key thinkers in the area of artificial, or machine, intelligence as well as being a serial entrepreneur and genius. Kurzweil believes we're likely to see smarter-than-human machine intelligences well before the end of the twenty-first century. He also believes that the line between humans and machines will blur as we use machine technologies to enhance our abilities, give our brains cognitive boosts and, among other things, upload the human consciousness to a hard disc. Ray talks of a "merger of our biological thinking and the existence of our technology, resulting in a world that is still human but transcends our biological roots." This merger "will allow us to overcome age-old human problems and vastly amplify human creativity. We will preserve and enhance the intelligence that evolution has bestowed on us while overcoming the profound limitations of biological evolution."[30]

The third front in the war against human limitation is perhaps the most important, and if the transhumanists were looking for an admiral of the fleet, then Professor Bostrom would certainly be a front runner. This is the "what the hell does it all mean and what should we do about it?" front, and Bostrom is well qualified to grapple with these big questions.

Arriving at Oxford I head for Broad Street and the Indian Institute, where Professor Bostrom has invited me to an afternoon seminar discussing transhumanist ethics. Inside I find a modern, plastic-chaired seminar room (a sharp contrast to the stonework Hindu demigods and tigers' heads outside). I'm the first to arrive but over the next fifteen minutes a cadre of serious-browed, sensibly shoed, studious-looking types drift in. Bostrom is one of the last to turn up and I have no time to make my introduction before the first presentation, given by Rob Sparrow. Sparrow is an academic I instantly warm too—he's an Australian bioethicist with a cheeky look in his eye, who's not ashamed to say that he is "genuinely confused" by the implications of his own work. Rob's worried that some thinkers in the transhumanist movement are ushering in a world of "free market eugenics."

Sparrow is particularly vexed by the principle of "procreative beneficence"—an idea advocated by Julian Savulescu, an Oxford contemporary of Bostrom's. The argument goes something like this: If after birth you willfully altered your child's genetic makeup to give her a medical condition that would reduce the quality of her life (say, asthma or Asperger's) this would be considered abuse, akin to beating your offspring or making them listen to songs by Andrew Lloyd Webber. Now consider a couple undergoing IVF (in vitro fertilization) who must select one of two embryos which are genetically identical except one has a defective gene that is likely to result in asthma. "Procreative beneficence" argues that if the technology is available to give parents this knowledge, choosing anything other than the "healthy" embryo is akin to abuse (although the principle still defends the right of parents to make the "wrong" choice).

Rob Sparrow doesn't like the implications of this kind of thinking, because I suspect he fears that, even with caveats, we are on dodgy moral ground. The phrase "thin end of the wedge" comes to mind. He asks us to imagine parents choosing between two embryos—one which has a condition that, on balance, means it will probably live five years less than the other. "Which embryo should they pick? Embryo A with the condition, Embryo B without the condition or should they choose the third option, chance, and toss a coin?" I am the only person in the room to choose tossing a coin because I figure quality of life matters more than the length of it. It simply turns out that the embryo with the shorter projected life span is a boy.

Soon we're considering the argument that the future is better off all female. Women live longer and tend to nurture offspring more diligently, women are less violent, and in such a world the shoe-making industry would never see recession. It's theoretically possible to manufacture sperm and egg from bone marrow, doing away with the biological necessity for Brad Pitt—and with a little bio-tinkering all women could be created lesbian, doing away with the desire for Pitt as well. But, of course, you could equally well argue that, with human-altering technologies, it might be just as reasonable to reduce men's proclivity to violence, extend their life spans and give them wombs. And what the shoe manufacturers might lose would more than be made up for by an increased demand for power tools. Sparrow cites the "gap" some men feel by not being able to bear children (I'm not one of them) and wonders, if there is a demand for male pregnancy, should it be an option freely available if the technology can be made safe?

I suddenly realize that for these assembled academics, the prospect of fundamental alterations of our biology (including radical life extension and human enhancement) is a *given*. They're not talking about what happens *if* it becomes possible, they're discussing what we might do *when* it's an option. And this worries many transhumanist critics. Sparrow later tells me, "The implications of taking transhumanism seriously are so radical and implausible that I think we should be much less inclined to do so."

The seminar ends, and as the attendees dissipate (several I notice lighting cigarettes as they leave the building), I introduce myself to my host.

Professor Bostrom is tall and slim with fine blond hair falling over a large forehead. His face is angular and inquisitive with deeply set eyes—giving the impression it has spent a good deal of time screwed up in concentration, which on reflection is probably just how the face of a philosopher should look. There's a kindness to his features—and a certain impishness there too. When he talks, he almost chews his words, delivering each with deliberate clarity—the native Swedish inflections giving his English an authoritative "weight." As we walk to his office, I ask Nick how many transhumanists like a smoke.

"They fall into two camps. Some are trying to stay as healthy as possible so they'll live long enough to get anti-aging therapies." He pauses. "Others believe that we're near enough to the technology to enjoy ourselves and not worry about the damage."

We turn down a nondescript side street and arrive at the Future of Humanity Institute, which is not nearly as grand as its title implies. In fact, the building we stop at could easily be an insurance company subbranch. The institute itself comprises just four modern offices and a meeting area on the second floor. You'll find none of the gothic splendor of All Souls College or the Bodleian Library at the Future of Humanity Institute. It's also remarkably hard to find a biscuit. In my mind's eye I'd imagined a plush wood-paneled office with a roaring fire where I'd be offered a fine malt whisky to accompany the learned discussion and then be taken for a feast at High Table involving a goose. The professor's office is disappointingly ordinary. It could just as easily be the haunt of a loss adjuster as that of a philosopher.

So, is transhumanism really "the world's most dangerous idea"? "It *is* a dangerous idea," he declares. "Once you start to modify our basic biology, if you go about it *unwisely*, the results could be . . ." He pauses. "Um, quite negative." But, he reasonably points out, "the bigger the potential risks are—of course, the bigger the potential upsides as well. So the key is to try and be wise and ethically responsible. Which is easier said than done."

He sees a radically different future shaped hugely by the advances of science and medicine. But I also get the impression he feels those advances are not being equaled by similar improvements in our ability to deal with the possible moral and ethical consequences. I suspect Professor Bostrom sees his role as trying to help our moral compass catch up. He explains this by referencing ants and Adolf Hitler. "Think of us like ants. Maybe you're a happy little ant, building a good ant hill. But maybe you're contributing to Hitler's war machine? I'm arguing it's better to wake up and try and understand the bigger context." I can't help it, I like the guy (Bostrom that is, not Hitler).

By his own admission this is something that human beings are "very bad at" and "often get wrong." He even suggests it may be beyond us, but this hasn't stopped him trying. "I want to think rationally about these things, rather than going for whatever makes the best story or gives you the most emotional comfort. That's a very big challenge."

It's important to remember that Nick comes from a very pro-transhumanist perspective. He is, after all, the joint founder of the World Transhumanist Association, which "advocates the ethical use of technology to expand human capacities" supporting "the development of and access to new technologies that enable everyone to enjoy better minds, better bodies and better lives."

I'm keen to understand where his optimism about transhumanist outcomes comes from, when so many people I mention the idea to visibly recoil. What drives him to bang the transhumanist drum in a world generally hostile to the concept? Throughout our talk, I pose this question in various guises, and Bostrom always answers in the same spirit—that, in his view, human life extension and enhancement will allow us not simply to live longer, but to *enjoy* living much more. When I ask, "What inspires and motivates you?" he cites his reading, his colleagues, but also states "I guess through feeling and experiencing something in this life and thinking 'why can't it always be as good as that?'"

This is the emotional driver at the core of the transhumanist dream.

"It's hard for a lot of people to see the problem—that life isn't always as wonderful as it could be—and part of the reason for that is our biology. Just like chimpanzees (or ants) we have biological constraints in the sort of thoughts we can have and the lives we can lead. To overcome this, the transhumanist project is developing technologies that can enable us to become all that we are in potential, by changing not just the world around us but human biology as well." Well, it's certainly more ambitious than getting a new kitchen. For him and many other transhumanists it's not just "better than human," it's also "have a better time than humans." And if you're going to have a better time, why stop at seventy or eighty years?

"Life extension is very central to me," he states. "First because for all the ordinary reasons we prefer to be alive rather than dead and, second, because if you think that the future might hold additional possibilities of radical enhancements of human capacity, the only way to get to experience that is to remain alive long enough, until those technologies have been developed."

Remember, when transhumanists talk of enhancement, they're going way beyond the capacity to switch on that fleeting "two pints in: brilliant at pool" ability at will. Transhumanists talk of boosting your intelligence, giving you Steve Austin's eyesight, access to a perfect memory or the ability to run faster than an Olympian. Again, all good for the individual, but is it good for our species as a whole?

In their paper "Protecting the Endangered Human," bioethicists George Annas, Lori Andrews and Rosario Isasi worry that:

The new species, or "posthuman," will likely view the old "normal" humans as inferior, even savages, and fit for slavery or slaughter. The normals, on the other hand, may see the posthumans as a threat and if they can, may engage in a preemptive strike by killing the posthumans before they themselves are killed or enslaved by them.[31]

They go on to argue that the potential for interspecies genocide makes the transhumanist agenda one of potential mass destruction. This is on the apocalyptic end of the criticisms of transhumanism, but it's all too possible to imagine a scenario where, because of cost or distribution or political agenda, some people get the opportunity to enhance themselves and lengthen their lives, while others don't. Isn't it very plausible that we could end up in a two-tier society of posthumans and unenhanced "normal" humans—with huge potential for conflict and disharmony, even war?

Bostrom argues that it doesn't necessarily follow that enhancement technologies will increase inequality. "So on the one hand people who are richer and better informed will presumably have better access to the latest technologies. But on the other it seems to be technologically easier to enhance the capacities of those who start off on a lower level. It just seems to be easier to fix a system that is already not functioning optimally. The further you are away from this optimum the more you start to benefit. This would tend to *reduce* inequalities, because it's just easier to lift up the floor than it is to raise the ceiling." In short, it's easier to help those of us with "suboptimal" genetics and health than to improve the prospects of those of us lucky to have already been dealt good cards. Looking at the link between income and life expectancy on an international scale and across time, Professor Bostrom's argument appears to be writ large.

In 1901, average life expectancy in the UK was forty-eight, and average income per person per year $6,264 measured in international dollars fixed at 2005 prices. (International dollars are a useful hypothetical unit of currency that measures value in terms of "purchasing power"—thereby allowing like-for-like comparisons of relative wealth across geographies and timescales.) A hundred and two years later, in 2003, the African state of Burundi achieved the life expectancy rates the UK enjoyed in 1901, but on an average yearly income of just $427 international dollars. Or put another way, Burundians are enjoying the same life expectancies of people who were nearly *fifteen times richer* than they are. It turns out that across the world the difference in longevity between rich and poor has narrowed and is continuing to do so.[32]

I challenge Nick with the opposite scenario—one where there is equal access to enhancement technologies. Is it possible this could slowly eradicate diversity? "For instance," I say, "we're very aware of homophobia. Could we 'enhance' somebody to become straight? So, rather than fight prejudice, why not just have a 'straightening' therapy?"

Then Bostrom says something that turns a key inside me. "I think we need to distinguish between good and bad forms of diversity. If nobody had lung cancer, there would be slightly less difference in the world. That's less diversity. But most people think it would be a good thing if we had less cancer."

This is my lightbulb moment when I "get" the transhumanist view of the future and why they think it's inevitable.

Ask someone "Do you think we should live forever?" and they'll probably say no. Ask the same person if we should continue to battle disease, they'll almost certainly say yes, me included. Transhumanism won't arrive in a revolution, it'll arrive one therapy at a time. Fading eyesight? Try large print. Now these spectacles. Now contact lenses. Now laser eye surgery. Now stem cell therapy to return your eye to its former youth. (In fact, Italian researchers have been using stem cells to cure corneal blindness for more than ten years.)[33] Bad hearing? Have a large cone to stick in your ear. Now a bulky hearing aid. Now an in-ear hearing bud. Now a cochlear implant . . . I realize, like it or not, we're already embracing the transhumanist project, we've just called it "medicine." We've inevitably blurred the line between "therapy" and "enhancement."

Just before the 2008 Beijing Olympics, South African Paralympic runner Oscar Pistorius (who runs on curved carbon-fiber "blades" and has been called "the fastest man on no legs") entered the competition's qualifying rounds, pitting himself against able-bodied athletes. He didn't quite make the grade, but he wasn't far off. Hugh Herr, himself a double amputee and bionics researcher at MIT, says, "It certainly will be possible in the future to build a limb that outperforms an intact limb, especially for running."[34] In fact, there is an ongoing argument among sports officials that Oscar's prosthetic feet give him an *unfair* advantage. It's conceivable that by the Olympics of 2012 or 2016, the "handicapped" will be unbeatable. In laboratories around the world, research is leaping ahead building limbs that are bonded to the body, controlled by the mind, stronger and lighter than the real thing, with artificial pressure-sensitive skin.

The Bionic Man isn't fictional anymore. Considering the growth in cosmetic surgery, it seems highly likely that in the not-too-distant future someone will elect to have a perfectly healthy limb deliberately removed and replaced with a "better" artificial one. Unlike natural evolution, which requires centuries to fumble an improvement, our man-made technologies quickly keep on doubling in power and halving in price, ushering in the prospect of the "disabled" outperforming those whom we now consider the able-bodied. Will I be buying enhanced cochlear implants in a few years the way I bought a mobile phone in the nineties? Will I buy a new body part grown from my own stem cells to keep me young? How far can this go? How long can I live? How enhanced could I be? Nick Bostrom suggests that the answers to these last two questions could respectively be "a very long time" and "as much as you like."

Take, for instance, the idea of "longevity escape velocity," seen by many transhumanists as one of the more convincing arguments for the prospect of immortality. This is the proposition that instead of our life expectancies gaining a quarter of a year for every year that passes, we reach a point where medical advances raise our life expectancy by *over* a year for every year that goes by.

Bostrom paints a mind-blowing scenario. "Once you have a sufficiently long life span then you would think that within that time there would be even more powerful technologies which can extend your life further. You can imagine that there will be ways of uploading into computers and then we'll have a space civilization where you could live for billions or trillions of years. So if you could survive for the next several hundred years (or it might just be the next fifty, we don't know), at that point your life expectancy might become literally astronomical and, yeah, that would be a profound change to the human condition."

Fly. Windshield. What does that kind of future *look like*?

"In my view it would be a distraction to focus on external things," he says. "It's better to think about what kind of mental state or mental activity would most correspond to what you would want to do if we had the ability to do whatever we wanted. So maybe try and find the highest reaches of human happiness and creativity and relationships—which will be refined even more in a posthuman future—but find the closest we can get today. What do you think the ideal is? Maybe the simple pleasures of family life, the ecstasies of artistic creation, romantic love, happy children eating an ice cream in a playground? I don't know. But whatever you think the ideal is, then one would think a posthuman condition would be able to realize that ideal and other ideals to a higher degree. That's the whole point of trying to get there—that it's something we think it's supremely worth getting to."

I'm struck by how very *human* Nick Bostrom's motivations are—on an emotional level they are the things most people care about: relationships, love, great music. I try to imagine myself "posthuman." Might I look back on this book with quaint amusement?—the disjointed ramblings of an un-enhanced mind, stalked by mortality—caught in his own transience, trying hard to understand "the big picture" but really one of Bostrom's ants, behind it all scared—literally—to death?

Short of an accident, I'm going to live longer than I thought. And I'll probably live healthier for a greater proportion of my life than my ancestors. I now believe that, at a minimum, stem cell therapies will become mainstream before I'm a pensioner—offering a very real possibility of repairing my aging human body, potentially pushing me into my second century.

Just a couple of months after my visit to Oxford, scientists at Imperial College London announce that by stimulating the bone marrow of mice, they can release a flood of adult tissue–specific stem cells into the bloodstream. No invasive surgery, no cultivating cells outside the body, no baby parts. Dr. Sarah Rankin, who led the research, likens the process to sending a fire engine to every street corner: "Once they're in the blood they're 'patrolling the streets.' Just as a fire engine passing a house on fire would naturally stop, likewise your stem cells will go to an area of damage."[35] Rankin estimates human trials will begin "in the next five to ten years."[36] Invasive surgery may become an increasingly rare occurrence. Is it possible that in the future, if I need a hip replacement I'll grow my own?*

---

*Researchers at the Columbia University Medical Center in New York have already demonstrated just such a technique in rabbits, using the animals' own stem cells to regrow joints. Elsewhere, Dr. Thomas Einhorn at Boston Medical Center has helped more than fifty patients with degenerative hip disease by using their own stem cells injected into the hip, to help generate new bone.

———————

That evening I visit the home of an old friend who lives in Oxford with her husband and their baby daughter.

One-year-old Abigail Billen is quite possibly a member of the last generation who will substantially worry about disease and aging (either that or they'll be ruled over by a race of genius-immortal mice). Their world is going to be a very different place from the one we live in today. At the very least it will need a new moral framework to deal with the results of biotechnology, new ethics of enhancement, new definitions of the word "therapeutic." She will almost certainly view her mortality quite differently from the way her father sees his.

"After you're forty, I find, all you really think about is death," her father says. "Your perspective completely changes. You wonder, 'Shall I bother to paint the bathroom door?—because I might be dead quite soon.' Before I was forty I never factored that in. But after forty I worried about it. It's just a background hum all the time." He's not joking, so I'm glad he at least bothered to cook the rather excellent casserole we're tucking into. Surely here is someone who might actively embrace the transhumanist dream? What if he could be freed of that "background hum"?

"Oh, I don't know," he says. "These people who want to live forever are tremendous egotists. Life is like a sonnet. It's fourteen lines and you know how to operate within its perimeter. It's the rules, isn't it?"

I am alive because the rules changed.

One Saturday at the start of the school summer holidays in 1981 my eldest brother was tasked with looking after me while our parents went shopping. He wasn't happy about this as he'd arranged to drive over to see his new girlfriend. Being a twenty-two-year-old man in the first flush of a young relationship, I suspect he had ambitions that didn't sit easily with having to take his ten-year-old sibling along with him.

We never made it.

Ironically, if Ian hadn't spent so long making sure my seat belt was a good fit, we would have missed the car that went straight through the Stop sign and plowed into our passenger side. The engine lifted off its mountings, twisted and landed on my legs, snapping both my femurs clean in half. My spectacles were found thirty feet down the road and the crash was heard a mile away. There was a lot of blood (mine).

My brother managed to crawl out across what was left of the bonnet but I was trapped in the wreckage. As I approached the crucial three-hour point of being pinned in, the paramedics decided the only way to save my life was to amputate both legs, which were on the brink of dying and taking me with them.

Then an ambulance man to whom I owe a great deal said, "If we damage his legs trying one last thing, he hasn't lost anything, has he?" Dosed high on morphine, I don't remember the fire engine and the ambulance attaching the ropes to either end of our fatally crumpled vehicle and literally pulling it apart.

I do remember being lifted out.

In the hospital, I received, I think, four pints of blood. Without it, the rescue effort would have been worthless. I didn't know it at the time, but I should have been thanking a nineteenth-century physician called James Blundell who performed the first successful human blood transfusion—and changed the rules about who lives and who dies.

I tell the story because I'm trying to imagine what would happen if any modern politician suggested we end the practice of blood transfusion. Medicine has been changing the rules ever since the first patient was saved, and we've welcomed it. Like it or not, we're not getting off this road, because we don't want to. Would any of us deny the newly wind-piped Claudia Castillo the right to go dancing, or run in the park with her children?

It's quite likely that I'll live to a hundred—and while that may not be long enough to get me into the running for becoming posthuman (although it might) it's longer than I was expecting and gives me a good frame of reference for my travels—to wit, what's going to happen in the next sixty years. It also means that if I have kids, I'll probably live long enough to see *their* children enter adulthood. And what my grandchildren might become is an interesting prospect.

Juan Enriquez, founding director of Harvard Business School's Life Sciences Project, argues that our change as a species will involve the "accumulation of small, useful improvements that eventually turn homo sapiens into a new hominid . . . a new species, one with extraordinary capabilities, a *homo evolutis* [*sic*]."[37]

Every advance in therapy is also a step toward the posthuman condition. The transhumanists know this. The rules *are* changing. Medical science is sounding the death knell of the old game. If we want, maybe we can fade the "background hum" to silence.

Aubrey de Grey has said his work isn't about keeping people alive longer per se but "to stop people getting sick." Radically longer lives are, he suggests, really just a "side effect" of good medicine.[38] Such future medicine, I'm beginning to understand, is based on the ability to understand and manipulate our genes and the cells that carry them—the oft heralded but poorly understood "biotech" revolution. To understand the future (and get a grip on whether the transhumanists really are onto something) I need to understand biotech. It's both a wonderful and frightening world. The next stage of my journey will take me not only out of the country, but way out of my comfort zone.

2

"The Most Wondrous Map"

Life is rather like a tin of sardines—we're all of us looking for the key.

—*Alan Bennett*

On June 26, 2000, approximately 22.5 kilograms of carbon, 70 kilograms of water, 2.5 kilograms of nitrogen, 1.4 kilograms of calcium, and a half-kilogram mix of fifty-four other elements (including just over a tenth of a milligram of uranium and enough phosphorus to make twenty-five hundred cigar-lighting matches) walked into the East Room of the White House.

"Nearly two centuries ago, in this room, on this floor, Thomas Jefferson and a trusted aide spread out a magnificent map," said then U.S. president Bill Clinton. The map in question was the result of the first coast-to-coast overland exploration of the United States and the "trusted aide" was Meriwether Lewis. But Bill had come to talk about another map: "Without a doubt the most important, most wondrous map ever produced by humankind." Was it true? Had they finally found a guide to the location of Rupert Murdoch's soul?

Seventeen years earlier, a man called George Church and a handful of his peers suggested making the map Bill was actually referring to: a completed "human genome"—a writing down of the genetic cookbook from which each human is uniquely baked. It's a cookbook small enough to be stored in nearly every one of your cells but contains enough information to explain why one bag of mostly oxygen, hydrogen and carbon is Bill Clinton, while a similar quantity of exactly the same stuff is you.

The accolade for sequencing the human genome was grudgingly shared by Celera Genomics (co-founded by Craig Venter) and a publicly funded international effort led by the National Human Genome Research Institute. Actually, Venter's team did it better, faster and over ten times cheaper.[1] They also tried to patent much of the knowledge.

Naughty Craig.

Subsequent outcry from press, public and many scientists, along with some hastily prepared legislation, put the kibosh on such capitalistic ambitions. Celera's stock price plummeted. Venter got fired and he ruefully remarked later, "My greatest success is that I managed to get hated by both worlds."[2] But he had helped give birth to a new era—one that he continues to play a pivotal role in shaping.

Even though you can't read the cookbook, your cells (and now technology) can—and getting hold of a copy (otherwise known as "having your genome sequenced") could one day play a part in saving your life. Beyond that, you may hardly have to visit the doctor, because you won't get ill in the first place.

Anyone on a quest to understand biotechnology at some point comes to the work of the same George Church who kicked off the human genome project all those years ago, a turtle-breeding, dyslexic polymath and now professor of genetics at Harvard Medical School. A man widely regarded as one of, if not *the* key, architect of the unfolding biotechnology revolution— according to *Wired* magazine a man so smart that "even the experts don't always get what he's talking about."[3] He hopes to usher in an era of "personal genomics"—where medicine and health care are tailored to the individual from the day you're born. To that end he's set in motion the audacious "Personal Genome Project" (PGP)—a study that will change medicine.

With some trepidation I phone the professor's number in Harvard and get straight through. "Sure," he says, in response to my request for an interview, but first he wants to test out my credentials. "Give me an example of something funny. You're a professional, right?" I suggest there's a certain irony in research that implies sufferers of Parkinson's disease could be helped by taking Ecstasy (usually a drug associated with *getting* people moving, not stopping them) and he seems satisfied.[4] "Have you heard about the people studying worm poop?"[5],* he asks, before inviting me to his laboratory.

Bingo.

But before flying to Boston, allow me to digress briefly to the picturesque university town of Tübingen, located twenty kilometers south of the Germany city of Stuttgart. My journey into the stuff of life starts, in many ways, here—for it was in Tübingen in 1879 that the most powerful alphabet known to humanity was discovered.

The man to write it down was one of the world's first biochemists, Albrecht Kossel. Except he didn't know he'd done it. He just knew he'd found four distinct substances within his colleague Johannes Friedrich Miescher's discovery, "nuclein"—a gloopy residue taken from the innermost part of human white blood cells. Miescher had extracted his nuclein from pus scraped off wounded soldiers' bandages. Kossel called the four substances "adenine," "thymine," "guanine" and "cytosine," and sometime later got a Nobel prize, in part, for doing so.

One letter (A, T, G or C) is all you need to represent each of the substances Kossel discovered. "A," for instance, is shorthand for the arrangement of five atoms of carbon, five of hydrogen and five of nitrogen that make up a molecule of adenine. Along with T(hymine), G(uanine) and C(ytosine) you have an alphabet of just four symbols—and with that alphabet, amazingly, you can write out a representation of the basic instructions to make a human being, your "genetic code." It is these four

---

*George is referring to the work of James H. Thomas of the Department of Genetics at the University of Washington, which aims to understand how different genes can lead to constipation. To this end he spends a lot of time watching nematode worms go to the toilet. I wonder how he answers the question "So! What do *you* do for a living?"

substances that nature uses to store all her recipes. My friend Mitch tells me, "I remember the four letters by reciting 'All That God Created.'"

It takes over three billion adenines, thymines, guanines and cytosines lined up in a row and in the right order to make the human recipebook as we know it—"the human genome"—a row of chemical letters you'll find in nearly every one of your trillions of cells (the most notable exception being the red blood cells). Written out as A's, T's, C's and G's on paper, it'd take up roughly the equivalent of two hundred volumes the size of the 1,000-page Manhattan phone book, and take nine and half years to read out loud (assuming you read ten letters a second and never slept, ate or went to the toilet while you were doing it).[6] That's roughly three gigabytes of data, or over twice the amount in an iTunes movie download of the genetic comedy *Twins* starring Danny DeVito and Arnold Schwarzenegger (which, by the way, is about as scientifically accurate as Darth Vader is cuddly).

Nature also provides a built-in backup of all that data. Adenine and thymine, and guanine and cytosine are like two couples who hang out but never swing. Adenine will only get into bed with thymine, and guanine only has eyes for cytosine. Within the confines of a cell, A's will snuggle up to T's and G's will soon form trysts with C's, one side reaching out with a "hand" made of hydrogen to cop a feel of the other's nitrogen or oxygen (this, appropriately, is called "hydrogen bonding"). The result? Two lines of data, one a "negative image" of the other. (Molecular biologists call the original line of molecules the "sense" strand, and the mirror image the "antisense" strand.)

If you add two long poles of sugar and salt for each line of molecular lovers to anchor themselves to (and so stand firm in their embraces) and give the whole shebang a twist, you'll end up with something that looks like a spiraling ladder, its rungs made of adenine molecules holding on to thymine molecules and guanine molecules holding on to cytosine molecules, like billions of trapeze artists frozen in midclasp.

There you are, then. The famous "double helix."

The most famous molecule in the world. A cell's individual hard drive. The code of life. A constantly referred-to book of recipes found in nearly every single cell that, amazingly, includes instructions to make everything *in* every cell. Your genome, encoded in the fêted and misunderstood "Deoxyribonucleic acid"—the bane of unwitting criminals who leave it behind at crime scenes, the premise for *Jurassic Park* and impossible to say properly after a pint of Greene King India Pale Ale.

It is a dead collection of atoms, bereft *of* life, but the cookbook *for* it (chemists call it "chemically inert"). No more life than a recipe is food, than a biography is a person or a page of musical notation is *All You Need Is Love.*

Salvador Dalí reportedly called it "a Jacob's ladder of genetic angels, the only structure linking man to God."[7]

While the data in your genome is different from that stored in the genome of, say, an African savannah elephant, both you and it share the use of this chemical alphabet. In fact, we not only share it with twelve thousand kilograms of unforgetting, floppy-eared *Loxodonta africana* but with every living thing on the planet—meaning the same four letters can be used to represent recipebooks for a whole menagerie of life and disease. This can lead to a worrying conclusion: If *all* life (plants, animals, bacteria, elephants, human beings, Robert Mugabe) uses this coding mechanism, this means *either* whenever life gets going it somehow evolves *exactly the same coding mechanism* (coming up with the ATGC alphabet independently) *or* we're all descended from a common ancestor. The first cell.

Robert Mugabe is family.

It is this last revelation that will make many an evolutionary biologist begin to see the wisdom in a third explanation—that God handed out the ATGC code to us all.

Wherever it came from, this is the code our cells read in much the same way components in your game console, deep down, read a code of electrical and magnetic charges represented by a two-symbol alphabet of 1's and 0's. When you think about it, that's a pretty arresting thought. We're used to putting code into a machine and getting out a game called "Sonic the Hedgehog." But the ATGC code can give you *an actual hedgehog.*

Why those instructions are arranged the way they are and what happens to them inside the cell is something Professor Church has been working to understand for most of his life. In doing so, he hopes to revolutionize medicine.

I arrive in America tired the night before Labor Day (like William Shatner, a Canadian import). The next morning I explore the good city of Boston as the day warms into a display of pure summery goodness. The season is going out on a high. The only thing that's bothering me is why everyone I speak to asks me where in Australia I'm from.

In the Public Gardens I come across a demonstration calling for health care reform. It's interesting to think that many of the medical conditions these demonstrators worry about could become a thing of the past if work like George Church's continues to bear fruit.

With the sun setting, I approach my hotel, passing the Broad Institute—a joint initiative by the mighty universities of Harvard and MIT. It is one of the many labs around the world spewing out documents full of A's, T's, G's and C's—labs that among them are writing down and cataloguing the DNA of pretty much anything they can get their hands on (including, in the Broad's case, that elephant I mentioned earlier). Currently, as a planet, we are collectively in the process of writing down the code of life in all its forms, including some that are extinct (their DNA being sifted out of hair, hooves and tusks held in museums).

The reasons for this gargantuan effort of genome librarianship are numerous but include, among other things, conservation. For instance, the University of Washington's Center for Conservation Biology have built "DNA maps" that highlight genetic differences between elephants from, say, Zambia and Malawi, helping them pinpoint where a particular specimen has come from. Using this information, they can work out the origins of illegal ivory—and in 2009 were able to direct law enforcement to the borders of Tanzania and Mozambique as a result.[8]

Elsewhere, the genetics research powerhouse Wellcome Trust Sanger Institute in Cambridge, England, is using genome sequencing to get new intelligence for another battle—that against cancer, taking the immortal words of China's most famous military strategist Sun Tzu to heart: "know your enemy and know yourself." Erin Pleasance and her team at the Sanger Institute examined entire genome sequences taken from two sets of cells from the same patient, one set ravaged by lung cancer (know your enemy) and the other as yet uninfected (know yourself). By comparing the two, they found that cells with the cancer had 22,190 DNA mutations (for example, a letter duplicated, deleted or replaced by the wrong one).[9] That averages out at more than one mutation of your code for every packet of cigarettes. (These changes occur because there are over sixty chemicals in your smokes that bind to, and mutate, your DNA.) That's still an infinitesimal percentage of the genome, but just like missing the letter "l" out of the word "public" in a government white paper, the result can sometimes be a whole host of trouble.

Webb Miller and Stephan Schuster of Penn State University are using a similar approach to protect the Tasmanian devil from a mysterious and contagious tumor that is endangering the species. "We are sequencing two specimens, one with the disease and another that seems immune, and hope to use the differences to guide a breeding program," says Miller.[10]

A final motivation for collecting all this genomic data is similar to the reason chefs buy lots of cookbooks—to add to their repertoire of dishes and get ideas for how to cook new ones.

The next day I head to the Department of Genetics at Harvard Medical School in the Fenway area of Boston. It's a slightly oppressive slab of glass and concrete that sits on, appropriately, Avenue Louis Pasteur— named after the iconic Frenchman who gave us pasteurized milk as well as the experiments that would prove to the world that diseases could be caused by "germs" (critters like the bacteria and viruses that George Church's work has helped unmask, derobe and force to give up their genetic secrets).

As I approach I'm assaulted by a plethora of signs that somberly and rather patronizingly announce "Let's Be Clear. Harvard Longwood Campus is 100% Smoke-Free both inside and outside." While there's mild irony in a medical school having to ban smoking, what really makes me chuckle is that alongside these words is *the world's most pointless map*. On it, campus buildings are colored dark blue, while the rest of the map is a lighter shade of the same color. The two-part key has a dark blue square, labeled "inside," and a lighter one, labeled "outside."

It worries me a little that an institution hosting some of the planet's smartest brains seems to think they need a diagram to help them work out if they're in a building or not—not that they can smoke in either location. I imagine it's the sort of thing that could make a logically minded scientist feel the need for a cigarette.

Professor George Church is a nice bloke. You'd want him at a barbecue. Words like "genial" and "gracious" spring to mind. When he talks, his tones are measured and gentle. That's not to say he won't challenge you or disagree about something, just that I get the impression that for George a disagreement is a chance to learn something, either about the subject under discussion or the person he's talking to. (During our initial phone call he'd almost warned me about this. "You'll find I'm a little *too* curious," he'd said.)

Physically he's a big chap. Six feet four inches tall and 245 pounds. That's slightly bigger than Muhammad Ali at his career heaviest, but George is almost the exact opposite. People often describe him as a teddy bear, and while that might seem overly cutesy, it's not far off the truth. He has a wide, open face, pale skin, a bushy uncle's beard and thoughtful eyes that seem to say, "Well, isn't this world interesting?" He works not because he's paid but because he enjoys discovering new things. "I have very expensive hobbies," he tells me. "Synthetic biology and personal genomics."

I can tell you a lot more about George you probably don't want to know. He has high cholesterol as a result of hyperlipidemia (for which he takes a drug called Lovastatin). Sometimes he suffers from an inflammation of the iris (a condition called, appropriately, iritis). He takes multivitamins as well as 100mg of "Coenzyme Q10" daily—a supplement that's been linked to lowering blood pressure, helping DNA resist breaks and lengthening the life span of rats. He's right-handed. His blood type is O positive. He's also a narcoleptic (a disorder characterized by excessive daytime sleepiness).

I don't know all this because a) I am inordinately nosey; b) George is a vocal hypochondriac; or c) we had the world's most boring conversation ever, but because Church has happily published this data, along with his genome, online for the world to see as "Participant #1" in what he describes as "possibly the broadest, most invasive study ever in genetics"— the aforementioned Personal Genome Project—a study that points to the future of my health care more clearly than any other to date.[11]

Over the coming years, the study will sequence a hundred thousand volunteers' genomes and try to correlate that data with their personal traits, their lifestyle and their medical history. It is the latest attempt to shed light on what is often oversimplified as the "nature *vs.* nurture" debate. After all, your genome is one thing, but you are another—as identical twins amply demonstrate.

Identical twins start out at conception with exactly the same DNA, and the correlation between their genome sequences remains as close as a gnat's whisker for the rest of their lives. But we all know that even identical twins can turn out differently. One twin leads a healthy life, exercises, doesn't smoke, eats a balanced diet. The other has a twenty-a-day habit, thinks whisky really is the water of life and eats fatty foods. You don't need a degree in medicine to spot which twin has which lifestyle. Same starting point, same code—different lifestyle, different outcome.

However, it's rare for twins to act as vastly differently as this, as two-thirds of the Bee Gees so amply confirmed, making it much harder to work out the subtle interplay between a Bee Gee's genome and his lifestyle. Is the ability to sing higher than a choirboy overdosing on helium a result of genetics, a diet rich in salami, long hours of practice in tight trousers, or a combination of all three? More seriously, what's the web of interactions between our genome and our day-to-day lives that triggers disease? Why is it that your sister always gets colds but you rarely do? Did Uncle Bob get heart disease because of one burger too many? Was it always written in his code? Or is he the victim of some yet unfound link between eating mustard sandwiches and working in a laundry? Why do some of us develop an allergy to latex? Why can a bee sting send skateboarding legend Andy Kessler into fatal cardiac arrest, but not me—and why is it the more times you're stung, the more likely you are to develop a potentially fatal allergy? Or to put it another way, how *does* Keith Richards stay alive?

At one end of the scale there are some conditions that are "in the genes." This is genetic determinism that's as unforgiving as Richard Dawkins in a bible class. Huntington's disease (which famously killed folk legend Woody Guthrie in 1967) is a good example. In most cases, sufferers will lose their coordination, memory and ability to concentrate—declining slowly to death. Changing your diet or environment won't help you in the long run. Currently, it's incurable.

Then there are conditions that are still unavoidable, genetically speaking, but you *can* do something about, lifestyle-wise. One is Følling's disease, also known as "phenylketonuria"—a disease which can amass too much of a substance called "phenylalanine" in the body, with devastating effects.

"Phenylketonuria," says George, proving he's a biologist by pronouncing the word without a flinch, "is something every newborn in a hospital in the developed world gets tested for. If you've got it, you have to keep away from phenylalanine and aspartame." That means steering clear of most artificial sweeteners, meat, poultry, fish or anything made with dairy products. If you don't, you'll get brain damage.

"It's hard, lifestyle-wise, but you can do it, and it changes the outcome from complete mental retardation to having normal brain development," explains George.

And, as we all know, personal lifestyle choices (beyond those forced on us by conditions like Følling's disease) can also affect your health. Ninety percent of lung cancers, for instance, are linked to smoking.[12]

However, things are rarely this simple.

It's important (if you're into curing disease) to know that the cookbook of your genome is split into twenty-three sets of paired volumes (one set from your mum, one from your dad) called "chromosomes." Within these volumes (a full set of which is kept in most of your cells) are recipes for cell components that your biological machinery constantly reads and then uses to make whatever those cells need.

The content of both sets is roughly the same but in any pair, one volume (the "mum volume" or the "dad volume") will often have a "preferred" recipe for a particular cellular component—this is the "dominant" version and is considered the definitive article in cell land. If the copy in the other volume is "recessive," it is generally ignored, referred to about as often as Saddam Hussein's *Little Book of Compassion*. However, you can have two copies of the same "dominant" recipe or two copies of the same "recessive" recipe (if you got the same data from mum and dad), in which case either volume can be referred to with equanimity. These paired recipes are called "alleles" (from the Greek *allelos*, meaning "other"), and the mix of recessive and dominant alleles (recipes) throughout your genome will largely determine the mix of your parents that is you: how tall you are, the color of your eyes, whether you can smell asparagus in your wee, whether you got mum's red hair, what diseases you're susceptible to and what you think of *The Late Late Show* (Okay, I'm joking about the last one).

Each recipe is generally called a "gene," although to be completely accurate a gene is *both* of the paired recipes ("gene" refers collectively to both "mum-given allele" and "dad-given allele") and it is the preferred allele that actually provides the instructions your cell uses. That recipe is usually for a "protein"—a protein being biologists' blanket term for a wide variety of cell parts.

Yes, on top of the lurid car-crash of syllables biologists so love when it comes to naming things they, like some politicians, can sometimes use a simple word you *do* recognize to mean something *entirely different*. When biologists say "protein" they're not referring to a nutrient found in meat, fish and eggs. For them "a protein" is a catchall way to describe a whole smörgåsbord of cell parts, a veritable RadioShack-esque catalogue of cellular bits and bobs.

Some proteins help us process food like the hormone insulin that plays a large part in turning sugars into cell fuel. Others manage chemical reactions like the enzyme telomerase that transhumanists get all excited about. (If you want a little more detail about why telomerase is causing such a fuss in transhumanist land, see Appendix: Damned Clever, Those Russians.)

A protein is the dish that gets cooked by following the recipe stored in an allele, chosen from a pair of recipe versions called a "gene," itself taken from the volume of a cookbook called a "chromosome," plucked from within the culinary *meisterwerk* that is your genome—a cookbook encoded in the four molecular letters of genetic code that we write out as A, T, G and C and stored on a hard disk called "DNA."

Having a particular arrangement of genes (or particular gene mutations) rarely *insists* you get a disease, it just changes the *chances* of you getting it—sometimes up, sometimes down. In most cases several genes have a role to play, all of them the genetic equivalent of a new card in a game of Texas Hold 'Em. Just as in poker, it's very rare that the cards you get will *absolutely* determine how the game will go; they just nudge the odds with varying degrees of force in a particular direction. Your environment and lifestyle are cards in the game too. Vegetarians who don't get enough calcium increase their *chances* of osteoporosis. That said, those who avoid meat *tend* to get colon cancer less because they don't eat animal fats.

It's the same with your physical traits. There are very few one-to-one relationships between specific genes and traits (dimpled cheeks and your blood type are two of only a few notable exceptions). Your height, for instance, is dependent on the interplay of several genes, and (again) your diet.

Understanding the steps in the genome/lifestyle dance is, to a large part, tied up in understanding how many times the cell is referencing different recipes in the genome—"gene expression" in the parlance of biologists (and surely a stage name waiting to be claimed?). If it's cooking up four hundred copies of recipe A for every three copies of recipe B, this can tell you something. It might tell you what you've been infected by and how badly. (This is because cells are easily distracted by foreign DNA that's been injected into them by viruses, a process George will later summarize as, "Oh, oh look over there! Naked DNA! I love naked DNA!") Or it might let you know if you're in the danger zone to suffer from a particular condition. For instance, "over expression" of a class of recipe-genes called "oncogenes" increases your chances of getting cancer. Oncogenes encourage cells to divide and proliferate—which is useful if you're a baby and want to get bigger, but less useful if you've got a tumor that wants to do the same.

If a gene is a recipe for a protein, then "gene expression" is the number of helpings—and just as is the case with food, too much or too little of something can be bad for you (as anyone with Følling's disease will tell you). Nature and nurture. Genes and "gene expression," entwined like a couple doing the tango, an intricate dance of life where you cannot understand or appreciate one without the other. (Science writer Natalie Angier neatly summarizes it by saying "you can't uncouple nature from nurture anymore than you can uncouple a rectangle's length from its breadth."[13])

Craig Venter told *Der Speigel* magazine that so far the Human Genome Project has produced "close to zero" medical benefits and that "we have, in truth, learned nothing from the genome other than probabilities."[14] This is why the *Personal* Genome Project is so important, hoping to expose some of the more subtle steps in the interplay of our genetics and our lifestyles, to master some of that complexity. George Church is less interested in the "really, really sick" because it's not obvious to him that "the mechanism by which you become really, really sick is relevant to most of the population." The PGP isn't really about saving the desperately ill (although it'll be useful in that battle too). It's about stopping most of us from ever getting that way.

It is arguably the most ambitious and most important medical study in history—one that could potentially inform the future treatment of thousands of conditions and reveal hitherto unknown knowledge about what makes the machine called *Homo sapiens* tick. The future of medicine, in terms of practical links between the *theory* of genetics and "on the ground" medical practice, is being born here at Harvard.

If you take part in the study (and George wants you to), be warned: Not only will you have key parts of your genome sequenced, you will also be measured in myriad ways (capturing your traits), asked a huge variety of deeply personal lifestyle questions and encouraged to divulge your entire medical record. You'll donate skin cells that George will reverse-engineer back to stem cells. And then you'll be asked to share all this with whoever wants it.

The reason is simple—the more people who have access to the full data, the greater chance there is of someone finding those links between our genetics and the lives we lead. The PGP will generate an enormous amount of information and George wants it interrogated as quickly as possible, by as many people as possible.

The danger of one institution (e.g., Harvard) keeping the data to itself is, believes George, the danger that new knowledge will, at best, take a long time to emerge or, at worst, be missed altogether. "You never know who's going to be the expert that figures something out," he says. "It could be some computer scientist in India, or some teacher in Bangladesh. I don't want to be the police officer that says: 'Oh, you can't solve this problem. You don't have the right approval.' The more people that are looking at it, the better."

Maybe there's a link between Church's iritis, his turtle-breeding and a particular gene? One more genomed-sequenced turtle-breeding iritis sufferer in the PGP may be all that's needed to find out—and a budding biological Einstein toiling in obscurity in, for example, Des Moines who now has, via the Internet, the keys to a treasure trove of data. But this level of data freedom is almost unheard of in medical research. As George says, "There's a lot about the PGP which is far from business as usual."

"One of the big things is finding people who have the same alleles," he explains, referring to those paired "dominant" or "recessive" recipes. "If you can find one case where someone is getting ill, but in the other they're not, then the difference is probably environmental. You can hunt down the environmental components and bingo! So, you change your lifestyle or your food and you get better."

In principle, that isn't any different to stopping smoking to avoid getting vascular disease or lung cancer, but again it's the more subtle interplays which the PGP could expose that George is after, unearthing a whole bunch of personalized advice to keep us healthy. "We can say 'these are the recommendations for the environments that would be good *for you*, and here are drugs that *you* should and shouldn't take. Oh, and here are drugs you shouldn't take in combination.'"

Talking of drugs, studies like the PGP could release an avalanche of lifesaving medicines that *already exist*, but aren't currently available. Right now the world is sitting on a wealth of useful drugs that never see the light of day. According to a 2001 study by the Tufts Center for the Study of Drug Development, just one in five thousand proposed medicines make it to being approved for human use by the U.S. Federal Drug Administration. Even those that get as far as human trials have a less than one in four chance of making the final journey to being available for prescription. The reason is that, although they may have benefits for many people, they can be dangerous, even fatal, to a small subset of others. Because we don't always know how to identify who will benefit and who will be at risk, drug approvals are necessarily cautious (with every new Vioxx *imbroglio** making the process more so). The economic impact of this is that the expense of researching and seeking approval for drugs that never see the light of day has to be covered in the prices of those that do reach the market. This means the price tag to a drug company for each successful drug easily approaches a billion dollars (and can sometimes be twice that), which is one reason why drugs can be expensive.[15] Personal genomics could change all that.

"Currently drugs are getting rejected because they fail in one percent of the population," says George. "You can use genome data to figure out that one percent and give the drug to the other ninety-nine percent." As well as saving millions of lives, this could potentially send the cost of drugs tumbling by allowing more drugs to hit the market.

Getting these benefits, however, requires accelerating the adoption of a key technology.

The personal genomics revolution depends on cheap, mass-market human genome sequencing.

"It's just really a matter of cost," says Church. "We should have been bringing the costs of sequencing genomes down to an affordable level, and that's what I've been working on for thirty-two years."

This ambition is why, for instance, George made the designs of his genome sequencing machine, the Polonator G.007 "open source." The machine severely undercut its competitors on cost/performance, so he was literally giving away a commercial advantage. There's nothing to stop someone else copying and selling an exact clone, or using George's designs to inform their own. For Church, though, that's not the point. The quicker genome sequencing becomes commonplace, the quicker lives will be

---

*Vioxx was a popular brand of the painkilling drug rofecoxib often prescribed to patients with "chronic pain" conditions such as arthritis. However, a study revealed that there was an increased risk (roughly doubling from a 0.75 percent risk to 1.5 percent) of heart problems if the drug was taken for eighteen months or more—and it was withdrawn under a heavy media spotlight.

saved. This is why you'll find George advising so many competing firms, even founding them. Money isn't his motivator. Speed is.

It's therefore heartening to find out that the cost of genome sequencing isn't so much falling as *plummeting*, showing the same sort of acceleration in price-to-performance we're now blasé about when it comes to computers. "Craig Venter's first complete human genome sequence cost a hundred million dollars, layered on top of the three billion spent by the competing publicly funded effort," says George. Yet, as I write, in mid-2010, a firm called Knome (cofounder, George Church) will charge you just under a hundred thousand dollars to do the same. Nine teams worldwide are currently competing for the Archon X Prize in genomics which will award ten million dollars to the first team "to sequence 100 human genomes within 10 days or less" at a cost of "no more than $10,000 per genome." (I don't have to tell you who's leading the team a lot of people have their money on.) But the ambitions of the prize may already seem out of date. Early in 2009, California-based Complete Genomics (yes, Church is an adviser) announced that it could sequence a complete human genome for just five thousand dollars and thinks it's on track to do it for a thousand (although there's no commercial offering as yet).[16] Already there's talk of the hundred-dollar genome.[17] Oh, wait, David Weitz of Harvard reckons he can do it for $30.[18] If genome sequencing carries on progressing as it has, the ten-dollar genome sequence will soon be with us and one-dollar genome not long after. "DNA sequencing costs were halving roughly once per eighteen months (like computer hardware) since the 1970s," says Church. "But over the past five years it's been halving every four months." (This is just one of a number of startling examples of something called the "Law of Accelerating Returns" that I will come across many times on my travels—and which has some interesting implications for us all.)

It's no exaggeration to say we're on the brink of a genome data explosion that will revolutionize medicine. The PGP is hugely ambitious in attempting to create a database of a hundred thousand, but imagine a world where we all have a document of our genomes in hand, or where the ability to sequence the genomes held in our individual cells and compare them (as the Sanger Institute's ground-breaking cancer study did) becomes common practice? This heralds diagnoses that go beyond "Mr. Stevenson, I'm sorry to tell you you have cancer" to "Mr. Stevenson, you'll be pleased to know that we've sequenced your cancer and therefore we have a personalized set of treatments for you." And hopefully I should be able to avoid many conditions altogether by preempting them.

"Imagine the day when you and your doctor sit down to review a copy of your own personal genome," says the blurb for the X Prize.[19] "This vital information about your biology will enable your physician to inform you of your disease susceptibilities, the best ways to keep yourself healthy and how to avoid or lessen the impact of future illness."

In the same chapter of *The Art of War* that talked of the power of personal knowledge in battle, Sun Tzu also wrote, "It is best to win without fighting." John Wayne might have said "Let's head them off at the pass."

This is where medicine is going.

If you'll excuse the pun, George's church is a broad one. Because he's not stopping there. He doesn't just want to *read* the code of life. He wants to *write* it too. And that has both hopeful and terrifying consequences, as I'm about to find out.

Appendix: Damned Clever, Those Russians

A bold man has luck in his train.

—*Danish proverb*

"Every time a cell divides you lose little bits of DNA called telomeres at the ends of your chromosomes," says my landlord Emmanuel, a man I've never met in person but who, by a stroke of inordinate transatlantic good luck, turns out to be an assistant professor at the very same Wistar Institute where Hayflick and Moorhead made their discovery of the Hayflick limit for cell division over fifty years ago.

"Telomerase addresses this problem by adding those short pieces of DNA back onto the end of the chromosomes," he tells me. "Think of telomeres like the plastic tips of shoelaces that protect the ends from fraying." This would be all well and good if the amount of telomeres that were lost when a cell divided were completely replaced in the new cells that were created. But they're not. In fact, every time a cell divides, the two cells that result have slightly fewer telomeres than the single cell they came from. In a sudden insight while waiting for a train on the Moscow metro, biologist Alexey Olovnikov worked out why this was, linked telomere shortening to aging and explained the cause of the Hayflick limit all at the same time. As James Bond might have said, "Damned clever, those Russians."

Olovnikov knew that as cells divide, their DNA must be copied into the two cells that are a result of that division. He began to imagine that the cellular machine responsible for copying the telomeres was a bit like a train passing over railway tracks. As the train rolls over the tracks, it makes a copy of the rails, but it needs to pass fully over a piece of track before it can copy it. At some point though the train hits the buffers at the end of the track. It can't pass over the last train-length section of rails underneath it, which means it can't copy them. The result is that the copy is one train-length shorter than the track it's just traveled on.

In cell land, telomerase is the engineer who comes and finishes the job off, marshaling telomeres together with a bunch of "telomere-binding proteins" (called "shelterin") to create the final cap that protects the ends of chromosomes. But our supply of telomerase runs down as we get older. The engineer hasn't got enough energy to do the job anymore until eventually he retires and "the telomeres get too short to protect the ends of the chromosomes," says Emmanuel. At this point "the cells turn themselves off and stop dividing to prevent genomic instability. The cell line begins to age and die." It turns out that the number of cell divisions it takes for engineer-telomerase to finally stop turning up for work is between forty and sixty—the Hayflick limit.

The argument goes that maybe we can continually replace that telomerase, keep the engineer healthy and thereby lengthen life. Work by Yousin Suh and colleagues "demonstrated that centenarians and their offspring maintain longer telomeres" compared with those of us who are shorter lived.[1] Her team also concluded that people with genes that give them an extra-active telomerase-making system tend to enjoy a longer old age. People who maintain longer telomeres are "spared age-related diseases such as cardiovascular disease and diabetes, which cause most deaths among elderly people."[2] (To be completely accurate there are many factors other than a lack of telomerase that contribute to telomere shortening, but that doesn't stop a boost in telomerase keeping our cells, whether healthy or cancerous, away from the Hayflick limit and "senescence.")

Emmanuel tells me there is now a general consensus that there is a *direct link* between senescence and telomeres. In fact, in May 2010 scientists working for the BioTime corporation reported that "by selecting for cells with sufficient levels of the immortalizing protein telomerase," they were able to "reset the clock of aging back to the embryonic state" and show that "time's arrow of development, as well as aging, could be reversed."[3] Like all companies with an eye on keeping their investors happy, the BioTime announcement was dressed nicely. This is early-stage research with an interesting result, but a long way from any kind of application. Nonetheless, it does hint that the transhumanists aren't as bonkers as some people make them out to be.

3

"My Lab Should Not Be Trusted."

> My momma always said, "Life was like a box of chocolates ...
> You never know what you're gonna get."
>
> —*Forrest Gump*

The biotech revolution goes way beyond the realm of human medicine—making it without doubt one of the most important stories of our age. Every living organism has a genome,* cookbooks we're now learning how to edit—the "genetic modification" that occupies so many headlines. The twin sibling of personal genomics is "synthetic biology"—the tools, techniques and knowledge that today allow us to recode cells and, in the future, will enable us to fabricate entirely new ones to our own designs.

Such things, not unreasonably, give a lot of people pause for thought. Fiddling with life is something many of us believe should be reserved for more divine engineers than ourselves, the argument being that biology isn't a tool to be reprogrammed in the same way we load software into our iPhones. Others echo the view of biologist (and codiscoverer of the structure of DNA) James Watson, who said, "If we don't play God, who will?"†

---

*So does every virus. In fact, the first genome to be sequenced was that of the snappily titled and bijou-genomed Bacteriophage MS2 virus, back in 1976.

†A brilliant biologist, Watson has lost some friends of late with overtly sexist and racist remarks. If anyone's going to play God, I hope it's not Watson.

One man—who, some critics argue, already thinks he *is* God—is Craig Venter. As the CEO and chairman of Synthetic Genomics he has stated (with only a smidgen of irony) that he has the "modest goal" of using synthetic biology to create new fuels and "replace the whole petrochemical industry." In May 2010, the papers announced he had "created artificial life." Venter himself was more measured. "We've created the first synthetic cell," he said. "We definitely have not created life from scratch, because we used a recipient cell to boot up the synthetic chromosome."[1] Venter made a synthetic copy of the genome of a bacterium. The DNA sequence (which included several nonfunctional DNA "watermarks" in order to distinguish it from the "natural" version) was written in a computer and then manufactured out of synthetic DNA. This was then implanted into a bacterial cell, which divided billions of times, each daughter cell inheriting the synthetic DNA sequence. The newspaper flurry of "Dr. God" headlines both overstated and overshadowed an important and symbolic milestone in synthetic biology. George Church tells me, "I'm a big believer in lots of milestones and celebrations along the way. I think we should celebrate it as the first fully synthetic genome functioning in a cell, because that's what it is." (He told *Nature* that "Printing out a copy of an ancient text isn't the same as understanding the language."[2])

It's with technology like this that Venter hopes to create genetically engineered bacteria that excrete diesel. He's not joking, and to prove it he did a deal with every environmentalist's favorite bad boy, Exxon Mobil, which is plowing hundreds of millions of dollars into the idea.[3] This kind of investment isn't an isolated event. San Francisco–based LS9 (cofounder: one George McDonald Church) have modified the genetics of *E. coli* bacteria so that if you feed them sugar, they produce fuels with "properties that are essentially indistinguishable from those of gasoline, diesel, and jet fuel"—attracting the attentions (and the money) of oil giant Chevron.[4]

This isn't science fiction. You can visit LS9's pilot plant in San Francisco, pour a gallon of their bacterially produced diesel into your car and drive off. Construction of a new factory with 2.5 million gallons of annual capacity is under way.[5] Want more? Venter has a life-form on the drawing board that makes fuel after eating carbon dioxide—a potential "one stop" solution to both the climate and energy crises. Elsewhere, Joule Biotechnologies is *already* using genetically altered photosynthesizing bacteria to take in waste CO_2, mix it with sunlight and make diesel and ethanol. (Joule's scientific adviser? Er, G. M. Church.) When *Newsweek* asked Venter if his fuels would work in today's cars and energy systems, he replied, "Basically everything we're making will work in the existing infrastructure."[6] He calls it "fourth generation biofuels." You call it gas.

Having just got my head around the idea that we can read our code (and the code of everything else), it now seems that there are plenty of people working on the idea of *rewriting* it—combining code from different organisms to make them do exciting and novel things. And with every organism on the planet being inexorably sequenced, there's one hell of a master cookbook of gene recipes to choose from and combine.

How about a bacterium that makes biodegradable plastic?[7] Or one that pumps out components for antimalarial drugs?[8] Well, both already exist (both versions of *E. coli* again*). If you're diabetic, it's almost certain your insulin supply is also produced by *E. coli* bacteria whose genome has been tinkered with, and soon genetically modified plants will be producing it too.[9]

It takes just a short leap of the imagination (and, some argue, the technology) to design and breed a mosquito that immunizes you against malaria rather than giving it to you (a bug you actually want to get bitten by) or to recode diseased cells to make them healthy (like rebooting or upgrading your computer). Venter's 2010 "synthetic cell" announcement

---

***E. coli*, I find out, is a family of bacteria that are a workhorse species in genetics. That's because they have a "simple," and therefore an easily fiddled-with, genetic makeup. Despite the bad reputation the bacteria get from a few black sheep in the family, it's generally a benign bug. You have a whole bunch in your gut right now generating Vitamin K_2 (a vitamin also found in lots of those überhealthy dark green vegetables like spinach).

foreshadows a world where we start to have the same control over biology that we have over software.

But as Franklin D. Roosevelt and Spider-Man both knew, with great power comes great responsibility.

Consider Huntington's disease. If we could recode carriers' DNA to save them from its (currently inevitable) consequences, should we? I know if I had the faulty gene in question, I'd take a DNA upgrade quicker than Ted Nugent gets erections at National Rifle Association rallies. And sufferers of Følling's disease would no doubt relish the chance of having some cheese on their crackers without, literally, losing their minds.

If we can recode people to avoid disease, we can probably recode them in other ways too. There are genes that have been linked to intelligence, athletic prowess, the ability to concentrate, aggression, even your propensity to vote. It's wise to point out that many of these links are often more media-made than scientifically proven (the furor over the "voting gene" being a key example), but there is a future coming where we could have the tools to nudge someone's abilities in one direction or another by fiddling with their code.

It's a tempting treasure chest—or Pandora's box—of genetic riches which opens up what ethicists might call "a huge can of (genetically modified) worms," amply demonstrated by the debate over what is known as "gene therapy."

To date, studies in "gene therapy" have tended to revolve around using viruses to deliver new or "good" code into our cells. After all, viruses work by importing their own DNA into cells, so why not engineer them to carry helpful genes rather than harmful ones? Well, as *Bad Science* author Ben Goldacre might say, "I think you'll find it's a bit more complicated than that." The actions of viruses aren't the easiest things to control or predict—as the tragic case of Jesse Gelsinger demonstrates.

Jesse had mild ornithine transcarbamylase deficiency, a liver condition which causes the body to stockpile ammonia and can easily lead to brain disease. Newborns with it generally slip into a coma within three days, suffer brain damage and die in infancy. Eighteen-year-old Jesse was "lucky" (in a relative sense) that his mild form of the condition could be controlled by diet. He was a willing guinea pig, understanding that the injection into his liver of an army of genetically engineered "adenovirus" wouldn't necessarily cure *his* condition, but could potentially help in the search for a genetic therapy that might one day be used to treat newborns and other sufferers. "What's the worst that can happen to me?" he said to a friend before heading to the hospital to take part in the trial. "I die, and it's for the babies."[10] Within four days his "worst that can happen" happened. His immune system took the virus to be hostile and went into overdrive, spiraling fatally out of control and resulting in multiple organ failure.[11] Jesse is now something of a macabre *cause célèbre* for those who question the wisdom of attempts to recode our cells.

And you can see their point. Other dangers associated with gene therapy include the risk that the code inserts itself into the genome in the wrong place and triggers cancer; that a friendly virus doesn't know when to quit and does "too much of a good thing"; or that a newly arrived gene gets "over expressed" with harmful results.[12]

It is over ten years since Jesse Gelsinger's death and gene therapy is still in the lab. Advocates point to successes like treatment for Leber's congenital amaurosis (LCA). Sufferers (who number about three thousand in the U.S.) lack a critical enzyme in the cells of the retina. As the years pass, their world gets darker and darker. In 2009, using an adeno-*associated* virus as a carrier (one that has much milder immune response than the straight-up adenovirus that killed Jesse), the same university that hosted the Gelsinger tragedy injected a gene that codes for the missing enzyme into nine-year-old Corey Hass's retinal cells.[13] "Corey doesn't use his white cane anymore," says his father, Ethan. "When he plays with his friends out back and it starts to get dark, he can still play out there and see. It's pretty amazing. They've changed my son's life forever."[14] There again, Jesse Gelsinger's father could say the same.

It's important to realize that Corey's condition involves a *single* faulty gene (related to a *single* enzyme), that physicians know exactly where to deliver the virus (the area affected is a tiny section of the eye) and that cells affected by LCA, while not doing their job, are rare in staying alive waiting to be fixed—hanging around like lumberjacks who've lost their axes. Viral approaches are still a long way from working to cure multiple-gene conditions that affect larger areas of the body. In short, viral gene therapy has had its "difficult second album" and is still waiting to see if it has a career.

In a paper published by George Church's lab shortly before our meeting, the situation is neatly summarized: "Our ability to sequence genomes," write the authors "has greatly outpaced our ability to *modify* genomes."[15]

But all that could be about to change.

"Delivering by virus is an *easy* approach, but it's probably not the *best* approach," declares Church. "The ideal approach would be to take the cells out, reprogram them, do some quality control, make sure you've changed the right genes and put those cells back into the body."

"Is that possible?" I ask.

"It's technically feasible. It's not super practical." He pauses. "Yet."

Professor Church's proposed "quality control" mechanism is the granddaddy of methods for sorting the wheat from the chaff.

Darwin's sieve. Evolution.

It won't rival Dan Brown for readership, but "Programming Cells by Multiplex Genome Engineering and Accelerated Evolution," a paper cowritten by George Church in 2009, could turn out to be one of the most far-reaching documents ever written.

In it the authors describe how, rather than trying to change a cell's genome one gene at a time, they introduced *multiple* genetic variations (e.g., an insertion, substitution or deletion of a chemical letter) into the cell's genome. Each variation was designed to nudge the cell's behavior in a desired direction (in the first instance the team wanted to see if they could increase production of lycopene, an antioxidant found most commonly in tomatoes). Over several generations of cells, a whopping twenty-four DNA modifications were introduced into the genome of *E. coli* cells.

Ever the engineer, George and his team built a machine to automate this entire process "for *large scale* programming and evolution of cells." Due to the method used to modify the DNA, different cells showed different variations to their genetic makeup, and each day the machine generated, wait for it, 4.3 *billion* variations of the *E. coli* genome. Within just three days of frenzied cellular reconfiguration, a small number of the nearly fifteen billion variants created were delivering the goods—a five hundred percent increase in lycopene production. Cultivate *these* cells a few billion times and "Job Done." It's not so much "survival of the fittest" as "survival of the cell that does more of what you want." LS9, George's biofuel company that wants to engineer *E. coli* to make gasoline, is now using the technique to improve the efficiency of their process. "What once took months now takes days," says Stephen del Cardayré, vice president of research and development.[16] A commercial version of the machine is in the offing. *Wired* magazine described it as a technique that "could make it as easy to rewrite a genome as to read it."[17]

Now we're cooking.

If you're treating a disease—say, of the liver—you could use a technique like this to quickly generate an army of "fixed" liver cells. "So it'd go something like this," says George. "You irradiate some chunk of your body, killing off diseased cells. Then you inject the engineered donor cells and they go straight to that depleted area." No DNA tinkering inside the body and no virus taxis. Instead, a batch of "tested for health" cells ready to put you right.

"The irradiation bit doesn't sound fun," I say. After all, irradiation is literally "exposing something to radiation," radiation that in this case is designed to kill tissue. It's something of a blunt instrument too. "It's not like you can pick off just the diseased cells, is it?" I ask. "You're frying some good stuff too?"

George laughs. "Yeah. Irradiating your body is not good for it, right? But you can imagine that with a little bit more synthetic biology, you could get these new cells to home in on where they're needed, and first *kill* whoever is there, but only kill them on a *need to kill* basis, right? You wouldn't need the irradiation."

It's a staggering proposition, and George says it so calmly that you'd be forgiven for not immediately grasping the importance of what he's saying. Like many engineers, George talks revolution as if he were fixing a plug, and I'm reminded of something I'd heard him say in a video of the 2009 Edge Foundation sponsored master-class, "A Short Course in Synthetic Genomics": "We can program these cells as if they were an extension of the computer."[18]

"The only thing that puts this kind of medicine far away is really *will*, right?" says George, dropping for a moment his genial, relaxed, rationalist demeanor, showing a heartfelt emotion. "The question is, how motivated are we?"

It's a hell of a question, because I suspect the answer is "a lot" and "no way, José" simultaneously. Because we're not just talking about curing disease. Sure, I'd welcome a cure for Huntington's disease, and giving sight to the formerly blind seems only fair for those who want it. But I'm less sure about built-in night vision for soldiers. Gasoline without digging up fossil fuels? Bring it on! Then again, surely there's a whole new class of biological weapons in the chest too?

Suddenly my mind conjures up a strange mix of *Dr. Strangelove* and *28 Days Later*. George is in the middle of a wasteland, decimated by rampant genetically engineered bugs that have destroyed the biosphere, leaving only him and the man who does movie voice-overs alive. "If only he'd known . . ." And the sun sets on a dead world.

I'm reminded of my chat with Nick Bostrom in Oxford. "Maybe you're a happy little ant, building a good ant hill. But maybe you're contributing to Hitler's war machine?" I say to George, "It seems to me that for every upside there's an equal risk. I mean there's some serious bioterror issues here, aren't there?"

"Oh, yeah," he says. "This is worse than chemical and nuclear weapons. Those weapons become less and less hazardous as they spread over larger and larger areas. But life? Well, a life-engineered weapon could become *more* hazardous as it spreads. You engineer the right virus, and infection can spread *exponentially*."

This is seriously dark. I've got a knot forming in my stomach.

"Surely, it only takes one bad bug to get out, and we could all suddenly be victims to a new disease?" I say.

"Yes, that's a totally fair scenario," says George as calmly as if he's just asked me if I want milk in my tea. I'm admiring his candor on the dangers of his work, but the knot is tightening, twisting.

"*Doesn't that keep you awake at night?*" I ask.

"I'm a narcoleptic. Nothing keeps me awake at night."

It's a nice moment of levity, which fades quickly.

"We will take some risks," says George. "Maybe even some existential risks."

"Existential risk," remember, is geek-speak for "The End of the World as We Know It." Bostrom explains it as risk "where an adverse outcome would either annihilate Earth-originating intelligent life or permanently and drastically curtail its potential." In summary: game over.

In the cheering paper "Existential Risks: Analyzing Human Extinction Scenarios and Related Hazards," written while he was at Yale University, Bostrom refers to some Australian gene fiddling undertaken in the search to create a new contraceptive. The result was a virus that killed every last one of those it infected and, worryingly, was highly resistant to vaccination.

It's tempting to claim that the search for a new Australian contraceptive was part of a larger program to engineer stay-cool beer and better cricketers, but it was in fact part of an effort to manufacture a pest control agent for rabbits. In the study, an extra gene was added into the mousepox virus. Despite its "pox" label, the untinkered-with version of the virus is associated with only mild symptoms. But with a single new gene added, it became a laboratory-wide rodent genocide. Within nine days, every mouse who'd been infected had made their representations to the great mouse-god in the sky. In the exact opposite of what happened to Jesse Gelsinger (whose immune system went into overdrive) the part of the mice's immune system that combatted viral infection simply shut down, leaving them defenseless.

In an interview for *New Scientist*, Ron Jackson, part of the team who ran the study, said that if "some idiot" put the equivalent human gene into human smallpox, "they'd increase the lethality quite dramatically." He continued: "I wouldn't be the one who'd want to do that experiment."[19]

What was doubly worrying about this experiment was that its results went against a general trend. It's pretty much the case that if you're altering the DNA of viruses and cells in order to make them do new things, you generally make them less good at the things they used to do. As George points out, "If I create an *E. coli* bug that makes gasoline, well gasoline is not good for most bugs, right? You can get them to survive but they're spending a huge fraction of their metabolism making gasoline rather than making baby *E. coli*. Compared to other *E. coli*, they're basically wimps."

But with George's new techniques for generating billions of genetic variants, the possibility of finding a super-dangerous and robust bug rises dramatically. (By the same token, of course, the search for the super-safe bug is also made much easier.)

A budding bioterrorist (or just naïve biohobbyist) needn't go to the bother of engineering a new bug either. You can buy a secondhand DNA synthesis machine from eBay, while the National Center for Biotechnology Information (NCBI), just outside Washington, D.C., has the A's, T's, C's and G's for the 1918 influenza virus (body count: between twenty and fifty million) written out and freely available online. (It's worth noting that this code has been used to resurrect the previously extinct virus for research purposes, and it's now alive and well in laboratories in the United States and Canada.) In fact, the NCBI has hundreds of flu genomes available to download—another example of the world's current genome sequencing frenzy.*

At the 2008 conference with the child-friendly title "Global Catastrophic Risks" arranged by Nick Bostrom, Dr. Ali Nouri from Princeton University's Program on Science and Global Security summarized it nicely: "What we worry about is that these tools will fall into the wrong hands and in some sense this is really an unprecedented threat—when a few people or even technically competent individuals gain access to tools that can produce so much damage."[20] In the book published as a result of the conference, Bostrom and Milan Ćirković wrote, "The technological barriers to the production of superbugs are being steadily lowered even as the biotechnological know-how and equipment diffuse ever more widely."[21]

The 2001 anthrax attacks in America or the 1995 sarin nerve gas attacks on the Tokyo subway act as powerful markers in the public consciousness when it comes to biosecurity. As the technology to manufacture pathogens gets cheaper, it's a natural worry that either "something will get out by accident" or that an ideology that sees mass murder as a legitimate action will gain the skills and technology to deliberately unleash a "terrorist pandemic."

Suddenly my future could be looking a lot shorter. One bad bug . . .

"Well, er . . . what do we do?" I ask.

What George says next surprises and heartens me.

"Well, we shouldn't depend on the goodwill of the scientists."

Good start.

"We should have *active* surveillance."

I lean forward. "You want to be watched?"

"My lab should not be trusted," he says.

"You're telling me you're untrustworthy?"

He smiles. "No, I think we're *very* trustworthy, but there's no reason to believe me, right? So, it's just easier to put us under surveillance."

---

*Since 1982, genome data held in the NCBI's "GenBank" has undergone exponential growth, doubling every eighteen months.

"And you'd welcome that?"

"Yeah, we're inviting that. And on top of surveillance, the entire bio-tech industry should be licensed too. At the moment, we're at a danger-ous situation because this stuff is easy to get hold of—and that has to change."

Well, I'm glad he sees it. George certainly isn't falling into the category of the scientist too blinded by the pursuit of knowledge to see a bigger ethical picture.

And, now I think about it, he's not the only one. Nick Bostrom not only arranges conferences to look at this kind of thing, but his first intro-duction was to invite me to a seminar on biotechnology ethics. Else-where, Stanford University's Drew Endy, another high-standing synthetic biologist, is well respected for bringing ethical considerations to the fore. "There is a lot at stake," he says in an interview for the *Technology Review*.[22] "I really want to see how far this field can go," but admits the answer will be "not very far" if the industry doesn't grapple with the bio-error/ bioterror issue comprehensively.

Before coming to Harvard I'd squeezed in a short trip to the Uni-versity of Southern Denmark to meet Mark Bedau, not only a founder of synthetic biology pioneers ProtoLife, but also a philosopher and an advocate for "scientific social responsibility."* "There are dangers, there are risks and you have to face them squarely," he said. "But the same is true with any other kind of powerful technology. This is why, from the very beginning, scientists in this field have taken the initiative to try and think through the implications, to think about how to do this in a way that is responsible. It's the scientists who are taking the lead in this field."

You'll find George arguing in print for "a code of ethics and stan-dards" and a "list of precautions only limited by our creativity" in an

---

*ProtoLife's approach to synthetic life is different to the Harvard model. While George is making artificial cell components, he's using nature's template. ProtoLife comes at the problem from a "first principles" standpoint, which says something like, "Okay, we're going to create life, it has a container, it has some code, it has a metabolism, but it doesn't necessarily have to be based on DNA, or pro-teins, or the rest."

article written for *Nature* four years before our meeting.[23] "Learning from gene therapy, we should imagine worst-case scenarios and protect against them," he writes—a reference to the Jesse Gelsinger tragedy. "This requires not just laws but monitoring compliance," he continues—something he believes can be largely automated. After all, you cannot make synthetic DNA without machines and machines could have surveillance and safeguards built in. By 2007, George was taking part in a joint study between the J. Craig Venter Institute, M.I.T and The Center for Strategic and International Studies along with a "Who's Who" of synthetic biology. The study outlined seventeen options for governance. "I've been advocating this surveillance strategy, this licensing strategy, since 2004," he tells me. (His "A Synthetic Biohazard Non-proliferation Proposal" was published in June of that year.)

"Can it be done?" I ask.

"Sure. Right now, there are bottlenecks which would be easy to license. So phosphoramidites are what you use to make synthetic DNA, right? You don't use phosphoramidites to make toothpaste, or lubricants, or anything like that. So if you license the few companies in the world that are making phosphoramidites then you're done, right? You make it hard to get a license and then you've a legal framework you can work with. Anybody making DNA using unlicensed phosphoramidites, or anybody who makes phosphoramidites without a license, is in trouble. Another area is licensing the machines that make DNA. It's not easy to make them. If you tried to buy the parts, that would set up red flags all by itself." (And presumably you'd have to stop people selling them on eBay too.)

"Like if I tried to buy some uranium?"

"Exactly. You want to make it as hard as possible to do this on your own. And we already have frameworks for this kind of thing—for making pharmaceuticals, for instance."

I'm heartened that the preeminent expert in the world of synthetic biology is also a strong advocate for oversight and legislation. Again, it's a long way from the popular cliché of the scientist who cares only for knowledge and has little time for ethics.

"I think there's general consensus that this can be done, it can be done fairly inexpensively." He pauses. "I realize this may not be an exciting answer . . ."

"But it's incredibly important," I say, relaxing a little.

"It's an example of pragmatic optimism," continues George. "Scientists and companies should be willing to accept licensing. You should have a license, same as a car. And if you license the whole ecosystem from the chemicals, the instruments, the know-how, the genes . . ."

As I write, a fairly bewildering array of consortia both public and private are in various stages of gestation, signaling the birth of the licensing regime George advocates. The U.S. National Science Advisory Board for Biosecurity is fostering international dialogue of biosafety issues, as are both the World Health Organization and the UN (the latter considering how the existing Biological Weapons Convention can be enhanced to encompass synthetic biology). The commercial syn-bio industry too has several overlapping initiatives—the International Association Synthetic Biology, the International Gene Synthesis Consortium and the Consortium for Polynucleotide Synthesis. In particular, members of these consortia are beginning to implement the scanning of DNA orders—making sure no one is ordering anything that looks like a pathogen.[24] It's heartening to note that in covering Bostrom's Global Stratospheric Risks conference, *New Scientist* chose to write an article entitled "The End of the World is Not Nigh."[25]

It strikes me that a race is on—between the science and our ability to implement a regime that will help us reap the benefits without unleashing a tirade of bio-horrors. Give us the medicine. Save us from the weapons. But if you can't save us from the weapons, are we sure the medicine is a price worth paying? Sadly there is no equivalent of "multiplex genome engineering and accelerated evolution" for policy makers—giving us four billion biosecurity frameworks to choose from by tomorrow. As ever, the science is moving faster than we are.

So, we'll have an accident. Synthetic biology will have its Three Mile Island, its Windscale, its Chernobyl. And somewhere, some ideology-driven buffoon will succeed in a deadly bio-attack giving us synthetic biology's 9/11. I suspect we'll have done most of the things we can think of to prevent it, but it'll happen anyway. We'll license the phosphoramidites. We'll put chips inside DNA machines that send e-mails to the FBI if they're asked to make anything dodgy. We'll update our weapons treaties.

But somewhere, something will go wrong.

It is now largely accepted that an overlooked piece of drain maintenance at the Pirbright animal disease research facility in Surrey, England, allowed an incidence of a 1967 virus no longer in circulation back into the UK countryside in the summer of 2007.[26] The resulting foot-and-mouth epidemic dealt a sucker punch to the British cattle farming industry. Our landscape seemed littered with pyres of beasts burned to contain the spread of the disease.

One. Bad. Drain.

As with the foot-and-mouth epidemic in the UK, we'll most likely contain future pathogen leaks to one degree or another. And then we'll learn how to make sure it's less likely to happen again, responding better and quicker, not least because we'll be able to sequence our enemy, find its weak spot and release countermeasures (chosen perhaps from a palette of fifteen billion pathogen-bashers made within days).

I can also foresee a further renaissance in medicine.

2002 figures from the World Health Organization show that cardiovascular, infectious or parasitic diseases and cancer account for sixty-five percent of deaths each year.[27] While it is not *impossible* to imagine an engineered pandemic virus suddenly taking out an equivalent thirty-five million annually, it is very hard to imagine this doing it year in, year out without some kind of fight-back from our medicine. A good example of this fight-back capability can be seen in our response to the threat from the H5N1 "bird flu" virus. (Bird flu is one of the "global catastrophic risks" that Bostrom's conference decided was more likely than others to put humanity in a tailspin.) Particularly worrying is the virus's ability to mutate (influenza viruses just love to change the rules of engagement). And just like an engineered pathogen a contagious version of bird flu would benefit from the increasing density and mobility of our populations, turning an outbreak into an epidemic at lightning speed—although new, fast, simple-to-use screening technologies like those being developed by the Institute of Bioengineering and Nanotechnology in Singapore could deliver mass disease screening for multiple conditions at airports, halting the spread of an epidemic in its tracks (their "Microkit" being just one of the fruits of the burgeoning nanotech revolution I find out about later on my travels).[29]

But now we have the ability to sequence and start to understand how the various viruses in the bird flu family work, allowing governments to stockpile the most appropriate antiviral drugs. This also gives us a far greater chance of developing preventive measures. We can "know our enemy" far quicker than we did in 1918. As of September 2009, the World Health Organization reported that there are no less than a hundred and ten potential vaccines in development to battle a bird flu pandemic.[28] During the 1918 flu pandemic there were none.[30]

In short, while we might have raised the chance of us engineering something nasty, we've more than compensated by developing techniques that could combat it. The same techniques will also help to see off things that *already* kill us today. Still, the accident will come. Some people will die, and synthetic biology will be the one found over the bodies with a smoking gun (or, more likely, a syringe). "But I was trying to save them," it'll say, and it will be telling the truth.

By the time I leave George Church, my mind is buzzing. *Everything* is changing. He has written: "We will learn so much more about ourselves and how we interact with our environment and our fellow humans. . . . I am optimistic that we will not be dehumanized but rehumanized, relieved of a few more ailments, better able to contemplate our place in the universe and transcend our brutal past."[31] This isn't like learning how to build a better internal combustion engine, or the invention of the TV. This is engineering *life itself*. It's tremendously exciting, offering hope on the one hand and potential terror on the other—just like a great movie. But how will it play out? Are we heading for a disaster, or a disease-free utopia? Probably neither. History is rarely like Hollywood. (If it were, we'd all be a lot thinner and less intelligent.)

Our biology is on the brink of moving at the speed of change we're used to seeing in computer hardware. I can almost feel the world shifting under my feet and, as I head out into the early dusk, something George said to me toward the end of our conversation begins to echo round my head.

"Evolution is far from over," he said. "It's accelerating, and something is going to replace *Homo sapiens*. Is it going to be wetware, or is it going to be silicon? Is it going to be genetically engineered, or is it going to be electrically engineered? Or, is it going to be some hybrid?"

There was a pause.

"It's just really a matter of when."

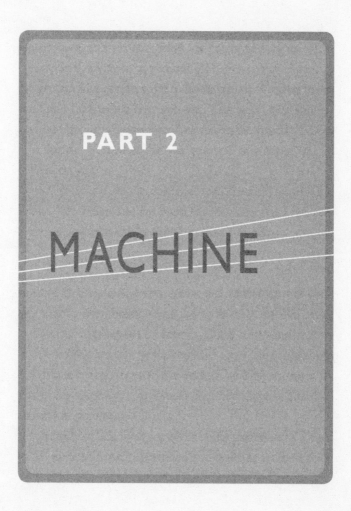

PART 2

MACHINE

4

"Can You Turn All the Buttons On?"

> Man has made many machines, complex and cunning, but which of them indeed rivals the workings of his heart?
>
> —*Pablo Casals*

Press the red button."

Leonardo looks down. It's the sort of big red button that wants to be pushed—the sort of button that launches devastating barrages of weaponry in old sci-fi movies. It's not just a button. It's a statement.

Gingerly he places his palm on top of it, blinking. He presses it. It lights up and (because this is modern-day Boston and not a scene from *Flash Gordon*) nobody gets vaporized. Satisfied, he stares wide-eyed at his teacher, Andrea, awaiting instruction. Leo doesn't know it, but he is about to prove he's a genius.

"Press the green button," she says. He obliges.

"Now all the buttons are on," she explains. Leo nods. His moment of genius is approaching.

Andrea turns both buttons off, and Leo blinks twice in clear surprise. *She just asked me to turn those buttons on!* As he glances down to take in this willful act of button extinguishing, she is putting something else in front of him.

"This is the blue button," she says. Leo regards the newcomer—the portent of his moment of glory. He moves his head from side to side searchingly. *Yes? What now?*

"Have you learned to turn all the buttons on?"

Ah! That game again. He nods confidently, reaches out to the blue button and, after a slight pause, pushes it down. She looks on impassively. Leo has a think.

And then it happens.

Tentatively he presses the red button. Then, finally he reaches out and presses the green button. He looks up with big adorable eyes, searching for approval. *All the buttons are on.*

"Very good," she says.

Leo has just proved he's a genius.

Now, as great achievements go, you may think this is hardly up there with Einstein's Theory of Relativity, the clavinet riff in "Superstition" or indeed that masterful goal by Pelé in the 1958 World Cup Final against Sweden, but what Leo has just done has some startling, possibly world-changing implications. Leo, you see, is a robot—and he's just done something robots aren't supposed to be able to do. *Think.* Or at least convince you that they're thinking.

In a 2008 article for *Scientific American*, Hans Moravec, research professor at the Robotics Institute of Carnegie Mellon University, reminded us that robotics had failed to live up to the predictions of the 1950s. Back then, faith in the calculational ability of computers led many experts to believe that with the right software "computers could become the artificial brains of sophisticated autonomous robots. Within a decade or two, they believed, such robots would be cleaning our floors, mowing our lawns and, in general, eliminating the drudgery from our lives."[1]

It's not the mechanics of making a body that's causing the problems, argues Moravec. While there have no doubt been many difficult challenges in replicating or making equivalents of human and animal body parts, these are largely being met, driven partially by the development of new materials and ever faster and smaller electronics. The Shadow Robot Company of London manufactures hands that demonstrate the same dexterity and range of motion as their human counterparts, with the strength to firmly grasp power tools and the delicacy to hold an egg. In 2005, Boston Dynamics revealed "BigDog"—essentially a four-legged robotic packhorse that can carry 340 pounds and traverse tricky terrain with apparent ease. Robotic vision systems can now capture images in resolutions that rival the human eye.[2] No, it's not the mechanics of the robot *body* that's the sticking point for the autonomous intelligent robot, it's giving it a *mind*. Which is why what Leo has just done is so important.

According to his creators at the Massachusetts Institute of Technology, Leo has just "generalized a task goal to a new configuration." That's geek-speak for "working out how to do something from having done something similar, but not identical, before" (a bit like being able to play Ping-Pong because you've played tennis, or being able to ignore tricky moral questions over dinner because you've worked in investment banking). It's a skill we learn as children and is essential for navigating the world around us—otherwise we'd find ourselves flummoxed by anything slightly different. Leo appeared to demonstrate that he could understand that "all" meant both the buttons in the first test, but *all three* buttons when the blue button was added. "All" is an abstract concept, which changes depending on circumstance. To you and me, the ability to "generalize" the concept of "all" seems easy—although this kind of abstract thinking is hard won in childhood. (My girlfriend, Caroline, remembers getting upset because her mother couldn't tell her exactly how many "a few" was.) Robots, like small children, find these general or abstract concepts difficult, which is one of the key reasons our sci-fi dreams of legions of smart machines happily helping us wash the dishes, refit space stations, catch criminals or perhaps take over the world seem a long way off. But if Leo is genuinely thinking, does this mean the intelligent robots are finally coming?

———

Leo doesn't look much like your average robot. If Yoda had slept with a German shepherd (the dog breed, not an *actual* German shepherd, you understand), the result would probably resemble this three-foot-tall ball of furry, fat-bellied cuteness. He has wide, expressive eyes, a small black nose and huge shaggy ears. He looks like a fantasy creature from a Spielberg movie, which isn't that surprising given that he was originally built by Stan Winston Studio, a creature workshop for films such as Spielberg's *Jurassic Park* and, aptly, the Kubrick/Spielberg *Artificial Intelligence: AI*.

The folks at Stan Winston Studio know how to make mechanical things look lifelike. But the animatronics in movies are human controlled. As complex as they are, as impressive and life*like* as they seem, they are superpuppets that "die" the moment their masters break for a biscuit. Leo is different. As soon as he arrived at the Personal Robots Group in 2002, his animatronics were stripped out and an autonomous robot was put in their place. Nobody was controlling Leonardo in the button exercise. He was acting alone. Leo, like you and me, is his own puppeteer.

As soon as George Church invited me to Boston, I knew I wanted to see Leo's "mother"—MIT roboticist professor Cynthia Breazeal, who, aged ten, saw *Star Wars* and dreamed that one day she might build something like C-3PO. Now, most ten-year-olds watching *Star Wars* want to be Luke Skywalker, Han Solo or Princess Leia. Those who instead see the hardware and think, "I'd like to make one of those," and forgo the lightsaber for the soldering iron, are just the sort of kids who end up at MIT.

Professor Breazeal is trying to build a robot we'd be happy to treat as a living (if not breathing) entity and not just an arrangement of servos, computers and sensors. She wants to create a personality we can interact and work with, even be friends with—to all intents and purposes, an intelligent, social, multipurpose machine. This scares the hell out of a lot of people. As Breazeal's former mentor, the restive, maverick genius

Rodney Brooks, writes in opening his book *Robot: The Future of Flesh and Machines*, "As these robots get smarter, some people worry about what will happen when they get really smart. Will they decide that we humans are useless and stupid and take over the world from us?" *,3

We're weaned on a diet of Hollywood movies that generally cast future robots as perilous weapons of war and aggression (*The Terminator*) or scheming murderers (like the iconic HAL in *2001: A Space Odyssey*), with only the occasional *Wall-E* as light relief. It's hardly surprising we fear being dominated by our mechanical progeny when our dreams are so dark. Noel Sharkey, professor of Artificial Intelligence and Robotics at the University of Sheffield, told *New Scientist*, "Isaac Asimov said that when he started writing about robots, the idea that robots were going to take over the world was the only story in town. Nobody wants to hear otherwise. I used to find when newspaper reporters called me and I said I didn't believe [that] AI or robots would take over the world, they would say "thank you very much," hang up and never report my comments."4 It's one of the reasons Cynthia likes to keep writers and TV crews out of her lab. Which is a problem for me. Because if anyone can give me a clear steer on the future of intelligent machines, it's Cynthia Breazeal.

Before leaving London I'd spent several enjoyable hours sparring on the phone with the formidable but hugely likable Polly Guggenheim, Breazeal's PA and gatekeeper. Polly had informed me that the Personal Robots Group has a policy that Cynthia gives one ("and only one") face-to-face interview a year. "We had the BBC here last year." A pause. "Never again." The message was clear: *I say no to people much more impressive than you.* "Who did I turn down last week? I've a list here somewhere. Would you like me to read it to you?" Polly speaks in a no-nonsense drawl built to

---

*Brooks's career, by his own admission in the book, has been one long exercise in upending accepted ideas: "I would look at how everyone else was tackling a certain problem and find the core central thing they all agreed on so much that they never talked about it. Then I would negate the central implicit belief and see where it led."

come out of the mouth of a New York policewoman detective. Still, after eight conversations she hadn't said no, which gave me the hope I might be in with a chance. (Polly doesn't give the impression of being the sort of person to dally with niceties once she's made up her mind.) It was agreed that Cynthia would quickly interview *me* over the phone, to see if I was worth bothering with. At that point I came straight out with it.

"Okay, seriously, what does it take to get this interview?"

"You really want to know?"

"Of course."

"Dark chocolate. *Seriously good quality* dark chocolate."

And so, after a brief chat with Cynthia, who sounded amiable and sparky, in which I promised to arrive at MIT bearing superior confectionery, I got my interview.

I find the Personal Robots Group on the fourth floor of the geek-famous MIT Media Lab, a boxy white structure on Ames Street that's beginning to look rather tired and shabby. Its residents will soon spread out into a shiny new ultramodern building next door, designed to encourage even more of the interdisciplinary collaboration that makes MIT such a hotbed of innovation. If there was a dial you could attach to a structure to measure the number of new ideas being generated inside, one attached to the Media Lab would probably blow a fuse before you could take a reading.

"One of the things that is so great about being at a place like MIT and the Media Lab," Cynthia tells me later, "is that we're not just a pure engineering group. We have skills and abilities that range from the aesthetic to the highly technical and scientific. That's unusual and it means we can answer questions that other labs find it hard to." It's a fair point. Where else, I wonder, will you find research groups with titles like "Affective Computing," "Biomechatonics," "Cognitive Machines," "eRationality," "Information Ecology," "Lifelong Kindergarten," "Smart Cities," "Synthetic Neurobiology" and, sharing the same lab space as Cynthia's "Personal Robots" group, "Opera of the Future" (that looks at "how musical composition, performance and instrumentation can lead to innovative forms of expression, learning, and health") all nestled up against one another?[5] It's bonkers, in all the right ways. It's also an unholy mess.

Cynthia's lab reminds me of the scene in *Blade Runner* when genetic designer J. F. Sebastian returns to his apartment—a cluttered, chaotic jumble of rooms, strewn with antiques and machine parts, and inhabited by comical pet creatures of his own design. As I enter, the plastic skeleton of a small robot sitting on a desk covered with computer parts raises its gaze from Ph.D. student Dan Stiehl to look at me. This is "Huggable" (without his fur)—a robot teddy made for small children and, Dan explains, designed to act either as a companion for when they are sick, or as an early education aid (where caregivers and teachers can control some of the robot's moves to encourage, for instance, the reading of a picture book). Even with all his mechanical guts exposed, Huggable gives the definite impression of being alive and I'm suddenly reminded of a passage from *Life on the Screen* by Professor Sherry Turkle. Turkle is another MIT-er with a particular interest in how people and computers interact and, as Brooks remarks in *his* book (before going on to quote the same passage), "something of a critic of the claims of artificial intelligence research," Turkle had visited the Brooks's lab and met "Cog"—one of the very first "sociable" robots made by MIT.

"Cog 'noticed' me soon after I entered its room. Its head turned to follow me and I was embarrassed to note that this made me happy," she wrote. Turkle recalls competing with another visitor for the robot's attention and behaving as if in the company of a sentient creature. I know what she means. I'm quietly pleased that Huggable has found me interesting. I'm reacting to him just as Turkle did to Cog—like he's a person. Huggable tilts his head up slightly to gaze at me square on, waggles his ears and extends his robot paw. I laugh, a little nervously. It's spooky, but in a nice way. The cutting edge of robotics seems about as far away from *The Terminator* as you can get.

Dan shows me a "Tofu" robot—an experiment to see how cartoon animation techniques might be used as a way for sociable robots to express themselves. Tofu is "a squash and stretch robot"—and this one is a furry ball about the size of a grapefruit with a small head dominated by wide eyes that dart comically from side to side and two feathery ears, along with a tiny orange beak that makes a cute chirping noise. Just like a cartoon character, Tofu can squish himself down and stretch himself out. He is laughably charming and if I were a child I would have immediately tried to take him home with me.

I'm eager to meet Leo, and Dan guides me behind a curtain. Suddenly I'm face-to-face with arguably the world's most lifelike machine. Except he's not very lifelike today. Leo is switched off. His head is slumped slightly forward and the box on which he sits has its front panel removed, revealing a myriad of strings and pulleys that control much of his movement. I also notice that Leo is wired up to sensors that hang from his curtained-off portion of the lab and, it turns out, no less than seventeen separate computers that process the information he takes in and generate his "thoughts." Leo is a prisoner of his limitations and, when you add in all those computers and sensors, actually much bigger than the cute pot-bellied furball I'd taken to be his entirety. It's disappointing for me because Leo loses some of his personhood and I'm a little crestfallen.

In person Cynthia Breazeal is like a shiny penny. She's forty-two, but looks at least ten years younger, a youthfulness no doubt related to her sporting prowess (an accomplished tennis player, she briefly considered taking up the game professionally). She manages to mix businesslike (a neat bob of a haircut) with childlike (her cheeks seem to permanently have a smile hanging off them, even when she's frowning). When she talks she uses her whole body. Her hands are constantly moving, as if literally grabbing concepts and shaping them in midair as she speaks. Her shoulders jog up and down, her head tilts with every new thought or idea, and she seems eager to laugh. People might argue about whether her machines are alive or not, but there's no debate about Cynthia. She's about the most *alive* person I think I've ever met. "These are *good*," she says tucking into the chocolates with a broad smile. "You should have one!"

I decline, wanting all that chocolatey goodwill going into our interview. Plus I know how much they cost, and I'm not sure my tastes are expensive enough to do them justice.

Polly Guggenheim interjects, "It'll do your endorphins good."

"Antioxidants! Antioxidants!" exclaims Cynthia.

"Exactly! It's like red wine!" says Polly as she departs.

With the benefits of chocolate and red wine firmly established I tell Cynthia how affected I was watching videos of Leo, especially in comparison to dozens of other robots I'd viewed for my research. Here was something of the future of robots long imagined by science-fiction writers but that I'd never seen for real. Leo *grabbed* me (and indeed all the friends I showed him to, some of whom refused to believe he wasn't a puppet). I believed he was thinking. The tilt of the head, the blinking, the shrugging of the shoulders are all social cues that I naturally understood and responded to, making me attribute emotional states and thought processes to him. I gave him a personality. In fact, I couldn't help myself, because this was a robot acting in a, well, recognizably human way (and therefore the exact opposite of Keanu Reeves). I'd had a similar reaction to Huggable only minutes earlier. What, I wonder, are Cynthia and her team doing right?

"I think it's because we put *people* at the core of what we're trying to do," explains Cynthia. "A lot of work in robotics is still very focused on technology, but in our lab we put these robots in front of real people so we can understand their impact. My group takes the relationship between social robots and people *seriously* and we're trying to design for *both sides*. It's not just about having the robot understand people, we're trying to make people *understand the robot* so you're naturally able to use your own way of thinking about the world to understand what must be going on in the . . ." She pauses. "Er, mind, so to speak, of the robot . . ."

"So to speak?"

"Well, it's a loaded term, isn't it?"

Two minutes in and we've already hit a key problem that's dogged

Artificial Intelligence research since day one, namely the intellectual quicksand around definitions of "mind" and "intelligence."

"Do you think Leonardo thinks?" I ask.

Cynthia does a neat sidestep. "I can ask a range of philosophers that question and they'll all come up with a different answer." But I push her. If it's robot-minds rather than robot-bodies that have been holding our machine brethren back, and if she believes Leo *is* genuinely thinking, that's profound.

"You're kind of avoiding the question," I say. She laughs and considers for a moment.

"What Leonardo is able to do is *aspects* of what I describe as thinking." And then she's off, a rapid-fire walkthrough of Leo's major subsystems, about how he takes in "perceptual information," how he has "internal states," and how his "cognitive-affective circuits" are "influencing internal actions," "propagating that activation" which "ultimately boils down to physical action in the world." She takes a breath. "I mean it's *all there* in some sense, but it's at a much simpler and rudimentary level, I think, than what's happening in animals and people." So, *is* Leo thinking? Well, sort of. A bit.

Part of the problem with this sort of question is that it's human beings that ask it, and we have a natural prejudice. We model the concept "thinking" on how *we* perceive ourselves thinking. Breazeal's quite clear that robot thinking *won't* be the same as human thinking, just as robot emotions won't be the same as human emotions. "When we talk about emotions, it's not about whether a robot *can* have emotions but what *are* robot emotions? So dogs have dog emotions, they don't have human emotions. But owners have beliefs about what their dog must understand about them, but they really don't *know*, right? They project their own emotions onto the dog!" She laughs. "*But* they see the dog's emotions as being genuine for what a dog is, and they see the relationship as being genuine, and that's the point. I want people to accept robots as genuine personalities, so they can have an authentic relationship with them." So, *does* Leo have emotions? Well, sort of. A bit.

"There's a lot of confusion in the media who often seem to think 'you're going to replace people, you're trying to make *people-equivalent* machines.' I'm like, 'No! These are *robots*.' But they misreport this stuff. They don't get it!" she says, bristling a little. "You sit there for hours and they don't get the nuance because they assume you're trying to build a human equivalent."

Cynthia's interest in building humans extends to her three sons, not into her laboratory.

The question of comparing human and artificial intelligences was addressed by computer science genius Edsger Dijkstra, who said that the question "is about as relevant as the question of whether submarines can swim"—a warning against trying to come up with a human-centric definition of "thinking."[6]

In *Robot* Rodney Brooks argues that trying to come up with concrete definitions of intelligence "is a daunting and most likely fruitless task." Instead he suggests a scale: "A person mostly seems more intelligent than a dog, and a dog more intelligent than a salamander, and a salamander more intelligent than an ant. And perhaps we would be willing to concede that an ant is more intelligent than a wind-up toy. Just what makes one of these creatures more intelligent than another, however, is very hard to quantify." In other words, it's all relative. Albert Einstein is smarter than Lassie who is smarter than Leo. Perhaps more important, Albert Einstein's thinking is *materially different* to that of a dog, whose thinking is materially different to that of a particular design of sociable robot. Cynthia makes the point that those different intelligences are a function of the bodies they're born into. "So the question is, what kinds of bodies are critical for developing what kinds of minds? Dolphin intelligence isn't human intelligence for a reason. Cat intelligence isn't dog intelligence for a reason. I would never say that what we're doing with these robots is somehow explaining what humans are really doing."

———

In 1997, IBM's computer Deep Blue played and beat chess grandmaster Garry Kasparov. This is seen, rightly, as a landmark moment in AI research, but it's part of an old AI tradition of trying to get machines to replicate highly intellectual "symbolic" reasoning (the sort you need to master algebra, for instance, and the kind of tasks revered, says Brooks, by "highly educated male scientists"). It's no coincidence that the frontiers of AI research seem to have been populated with computers that get ever better at playing chess. But the big elephant in the room for this kind of AI research is that *Elephants Don't Play Chess* (this being the title of a paper Brooks wrote in 1990), but they do do any number of "common-sense" things that machines can't—as can dogs, and cats, and preschool children.* Sure, Deep Blue can occasionally beat grandmasters (and no doubt elephants) at chess, but it'll never be able to find shade or express its political views on Vladimir Putin. Deep Blue is "intelligent" in the same way a sledgehammer is a good tool for opening a packet of crisps, or a submarine is a good way of getting from one end of the local pool to the other. AI research *has* produced impressive results, if you view them in the right way. It's just that its successes are found in arenas where the situation is controlled or constrained in some way (you play chess and that's *all* you do). As Brooks points out:

> There's this stupid myth out there that AI has failed, but AI is everywhere around you every second of the day. People just don't notice it. You've got AI systems in cars, tuning the parameters of the fuel injection systems. When you land in an airplane, your gate gets chosen by an AI scheduling system. Every time you play a video game, you're playing against an AI system.[7]

So, does Deep Blue think? Well, sort of. A bit.

But we still haven't got anything like *Star Wars'* C-3PO, the multipurpose

———

*I like to imagine Brooks breaking into IBM HQ, whisking *Deep Blue* away to the African savannah and saying to it, "Go on then, find a water hole."

smart machine that we consider a real personality. Steven Pinker, Johnstone Professor of Psychology at Harvard University, has written:

> The main lesson of thirty-five years of AI research is that the hard prob-
> lems are easy and the easy problems are hard. The mental abilities of a
> four-year-old that we take for granted—recognising a face, lifting a pencil,
> walking across a room, answering a question—in fact solve some of the
> hardest engineering problems ever conceived.[8]

Brooks argues that "simple things to do with perception and mobility [are] a *necessary basis* for higher-level intellect."[9] In other words, if you
want to build something truly smart it needs to be in the world, and not
chained to a chessboard. (Garry Kasparov's ability to play chess, after
all, isn't isolated from the rest of Garry Kasparov.) In the parlance of
Brooks and co., it needs to be "embodied" and "situated," meaning it
needs a physical body via which it senses the world, and it is this, more
than abstract concepts, that should drive its actions.

"Sense and action. This is all I would build, and completely leave out
what traditionally was thought of as the intelligence of an artificial intelligence," wrote Brooks in *Robot*. Using this approach he soon began to
achieve impressive, lifelike results from his machines, much to the cha-
grin of the AI old guard. At the Second International Symposium of
Robotics, held just north of Paris in 1985, Brooks demonstrated Allen,
a robot that could autonomously navigate itself down corridors, avoid
pedestrians and choose a route based on the obstacles before it—all
using a "sense and react" approach. This was a marked contrast to many
previous attempts at robot locomotion that had often relied on robots
trying to build coherent internal "mental maps" of their surroundings
and referring to these internal representations, rather than directly to the
world. But time and again these "symbolic" models couldn't keep up with
an ever-changing world. Brooks recalls how strong shadows had foxed
Hans Moravec's attempts to achieve autonomous motion in an MIT robot
called, unflatteringly, "the Cart." "The Cart got very confused about its

model of the world, and Hans had to run out and fix the world occasionally so that the internal model was a little more accurate—it was much easier to change the real world than to change the Cart's internal model of that world." Allen by comparison was, as Brooks puts it, "glorious."

By "eliminating the process of reasoning" and making a machine "completely based on unthinking activity" that made a "direct connection of perception to action," he had built something that seemed to act much *more* intelligently. In animals, "complex capabilities are built on top of simpler capabilities," he argued—and started building his robots accordingly. It was a view that seemed way too, well, *simple* for those "highly educated male scientists," and Brooks was soon working in his favorite place—on the fringes of respectability.

Step forward two decades and Brooks has largely won the debate. You can find footage of the aforementioned "BigDog" on YouTube traversing inclines covered in soft soil or woodland debris, snow-covered slopes, even piles of rubble. There are two particularly startling moments in these videos. In one, BigDog begins to slip on some ice but is able to recover and remain upright. In another, it's given an almighty kick from one side. BigDog staggers but quickly regains its footing. It's not chess, but suddenly this machine looks *conscious*. It's Brooks's "sense and action" argument taken to a whole new level.

Cynthia's work carries on in the Brooks tradition. Step one: build a body. Step two: make it sense and react to the world around it. But it's step three where Cynthia comes into her own. *Make it social.* Or to put it another way, make some of its reactions "social" reactions. So, for example, if someone gets too close to the robot, not only does it back away, it might express its annoyance at having its personal space invaded with a grimace or sound of disapproval. If someone shouts at the robot, make it respond as if it's upset or sad. In these examples the robot responds directly to something you *do*. But Cynthia is striving for a robotic social intelligence that takes things into a whole new realm, responding to something you *think*.

"One of the most significant abilities we have been working on is the ability for Leo to reason about *your* thinking in *relation* to its own." In other words she doesn't want Leo to just sense and then react to what you *do*, she wants him to sense your actions, interpret them in a way that allows him to infer what you're thinking, and then react to that. In controlled experiments (with Leo wired up to all manner of additional sensors) his "brain" has been able to pass something that looks like what psychologists call "the false belief test"—essentially, being able to understand that *someone else* may believe something that *you* know isn't true. For example, you might know that Mother has moved a pack of cookies from one tin to another while your brother was out of the room. When he returns, your brother has an out-of-date belief about the location of the cookies. Crucially, you know *he thinks* they're in the first tin, even though they're not. His mental state (cookie-wise) is different to yours. You have a "theory" of his mind in relation your own. He is thinking differently from you about the same situation and you understand that. "No robot to date possesses full human theory of mind capabilities," says Cynthia—but under certain highly controlled conditions, she asserts, Leo can infer a limited subset of others' mental states and reason about them, prompting him to action. He does this by simulating what his own actions would be in a given situation. Cynthia calls this "self as simulator," which is a roboticist's way of describing what you and I do when we ask ourselves, "What would I do if I were them?"* Achieving a "theory of mind" for robots "is important if you want robots that are competent collaborators," says Cynthia, but admits we're still a long way off anything that looks like it in machines.

Having the robot understand you, your actions and your mental states is only one half of the interaction. As Cynthia said, it's not just about having the robot understand people, it's helping *people* understand the *robot*. "It's about getting robots to understand, connect and relate their mental

---

*It's something you and I can do easily. If we couldn't, we'd never enjoy a film (putting ourselves into the minds of the characters) or be able to cross a road (by guessing at the thoughts of approaching car drivers).

state and yours—so you can work together in a normal, social way," says Cynthia, "You 'get' each other." This is because the Personal Robotics Group is trying to understand not only how to design a robot that can learn, but how you design a robot that's *teachable* by a person who knows nothing about robotics, AI or machine learning. "You look at the research in the field and the people who are training the robot are often the *same* people who *designed* the robot and who have an intimate knowledge of how to interact with it to make it work," says Cynthia. "The challenge we're addressing is, how do you allow Grandmother to teach the robot? The world is full of people, so social forms of learning—observation, imitation, tutelage—all these things are really important. We're probably one of the few groups in the world that has robots learning from people who know nothing about robotics." She laughs. "And I've gotta tell you, it's not easy to design a robot that can learn from people, because people do all kinds of stuff!"

Of course, as humans we learn to deal with all the "stuff" that other people do as we are growing up, and are hopefully nurtured by our parents and caregivers. Children generally develop in a safe and constrained environment where they can make mistakes and their brains can mature. This is all part of becoming intelligent in a way that means you can read a book, do your job, realize if someone is upset, drive a car and maybe (just maybe) understand cricket. After all, Garry Kasparov didn't pop out of the womb with a detailed knowledge of the Petrov Defense. First he had to learn how to walk, talk, eat and "generalize task goals to new configurations" just like the rest of us.

"Look at how often children fall when they're learning to walk," says Breazeal. "I mean, it's like *huge*." Alongside the physical tests are a whole load of abstract mental ones that children also spend a long time failing to grasp. Take for instance the ability to understand humor. "My kids are at that great age where they don't quite get jokes yet," laughs Cynthia. "I mean, they're really bright little guys at four and five, but it takes a long time to get the kind of social understanding that lets you know if someone is joking or not, or why something is funny. It's not until you start really digging into all this that you start to realize how deep and profound and *difficult* this kind of stuff is to do. Then you start to appreciate how amazing it is that *people* can actually do it!" She's laughing a lot now. "You just take it for granted that a smile can be interpreted as bemusement, or joy, or any number of things. It's obvious. You just get it." She raises her eyes and holds open her arms in an expression that (rather demonstrating her point) I can instantly interpret. *Holy moly!*

"Now!" exclaims Cynthia, "you try to build a robot that can do that. *Oh my God!*"

Take computed attempts to capture the nuance of "natural language" (the ability to use words in the way a human might, in a conversational rather than "formal" style). One popular example is Eliza, developed by MIT's Joe Weizenbaum in 1965. (Eliza was named after Eliza Doolittle from the George Bernard Shaw play *Pygmalion,* in which professor of phonetics Henry Higgins teaches the East End flower girl to imitate the speech patterns of high society.) Weizenbaum tried to get Eliza to mimic a technique used by many psychotherapists—reflection. Reflection is essentially a "mirroring" or restating of what has just been said and lets clients know they are heard, hopefully encouraging them to open up more. So, if you say, "I want to kill my husband!" they might respond by saying something like, "Your husband is annoying you?"

You can find various copies of Eliza on the Web. It's surprisingly good. In fact, it compares favorably to other natural language programs developed much later (highlighting again how hard it can be to improve the performance of AI in areas that humans take for granted). As part of my research I decide to get some online therapy.

I start off by typing, "I'm writing a book."

"How long have you been writing a book?" inquires Eliza.

"A few months," I respond.

"Tell me more," she counters, as if genuinely interested.

"I'm trying to work out what the future might be like."

Then she gets heavy.

"Do you believe it is normal to be trying to work out what the future might be like?"

I have to say this makes me think. Eliza really is turning out to be like a therapist. But then it all starts to go, well, *a bit weird.* After due consideration I conclude that yes, actually it is perfectly normal to try and work out what the future might be like and reply simply: "Yes."

Eliza's response?

"I see."

Even though Eliza's a piece of code running in my browser, this still feels strangely like an awkward silence. I'm uncomfortable.

"You don't do well at parties, do you?" I type.

"Oh . . . don't do well at parties do me?" says Eliza, and I'm not entirely sure what she means.

"Are you asking me to do you?" I reply. "Maybe you do better at parties than I thought!"

"Why are you interested in whether or not I am asking you to do me?" she says. "Maybe me do better at parties than you thought?"

I can't work out if Eliza thinks I'm a pervert or is a machine impersonation of a nymphomaniac foreigner. Poor Eliza, she failed to get the irony in "You don't do well at parties, do you?" which led us somehow into a dodgy porno/comedy script from 1976.*

It's results like this which lead AI and robotics researchers to suggest that the best approach to building intelligence is not to try and make an "off the shelf" mind, but to grow one in the robot, to bring it up. Children slowly learn not to make mistakes like Eliza's over years and years of interacting with those around them. "Natural language" skill takes *thousands of hours* to master (and we can all still get it wrong, as any "misunderestimated" president will tell you).

Perhaps if AI research had taken note of one of its founding fathers,

---

*Later I will meet Ray Kurzweil and remark on the poor state of natural language programs. He suggests I try Kurzweil Technology's Ramona, which he hopes will be better. It isn't.

it may have saved some time. In a landmark and it turns out startlingly prescient paper, written in 1950 (in which the first line was aptly, "I propose to consider the question, 'Can machines think?'"), the great, tragic, war-shortening Alan Turing* wrote, "Instead of trying to produce a program to simulate the adult mind, why not rather try to produce one which simulates the child's? If this were then subjected to an appropriate course of education one would obtain the adult brain."[10]

This is Breazeal's approach (with the caveat that her adults will think like robot adults, not human adults). For her the relationship between a child and its caregivers gives crucial insights into the development of intelligence. Indeed, numerous studies have shown that a lack of care, interaction and contact in human childhood directly affects brain development, the most memorable and horrific examples being of Romanian orphans hidden away by the Ceaușescu regime. Robots, she argues, need parents too.

If Cynthia is right, and we need to raise our robots just as we do our children, you might think truly intelligent robots will necessarily be few and far between. But the thing about robots is you can replicate them. Once we've got *one* intelligent robot brain, we can copy it to another machine, and another, and another. "One" quickly becomes "many" and the robots have finally arrived, bringing an explosion of "strong AI" (what Ray Kurzweil describes as an "aptitude resembling the broad, deep and subtle features of human intelligence, particularly its powers of pattern recognition and command of language"[11]). And of course, it may

---

*Turing invented and built some of the first programmable computers, which were used to crack the previously "unbreakable" Nazi Enigma codes in World War II. However, while the British government was keen to accept his genius, they were less understanding of his sexuality (homosexuality remained illegal in the UK until 1967). Following his admission of a relationship with another man, he was charged with public indecency and offered the choice of either imprisonment or chemical castration. He chose the latter. It is largely believed that this experience and those that followed led Turing to commit suicide on June 7, 1954. In 2009, the UK government apologized. In a statement released by Downing Street, Prime Minister Gordon Brown wrote, "his treatment was of course utterly unfair and I am pleased to have the chance to say how deeply sorry I and we all are for what happened to him. . . . So on behalf of the British government, and all those who live freely thanks to Alan's work I am very proud to say: we're sorry, you deserved so much better." Many believe Turing's work cut the war short by two years. He really did deserve much better.

not just be us (the humans) doing the copying, it might be the robots themselves. It takes a lifetime to build a human brain and (hopefully) fill it with wisdom, but copies of robot brains can pour off the factory floor. And because technology improves at a startling rate (technology generations delivering improvements faster than biological evolution ever can), one has to consider the possibility that things won't stop there. Once we achieve a robot with human-level (if not humanlike) intelligence, it won't be very long until robot cognition outstrips the human mind—marrying the "broad, deep and subtle features of human intelligence" with instant recall, flawless memory and the number-crunching ability of Deep Blue.

Cynthia may want to build a human-robot team, but what happens when one half of that team (the human) is starting to look like the village idiot? This is what has been called "The Singularity"—the moment where the generation of new knowledge (through a merger of humanlike talents of imagination, curiosity and creativity with machinelike computational muscle) starts to look like a rocket leaving a launch pad. According to The Singularity's prophets (most notably Ray Kurzweil), when this happens there is only one strategy open to us: We must merge with our machines.

If you can't beat 'em, join 'em.

The problem with all this for a lot of people is that it seems, well, *a bit unfair.* Many can't believe God's plan for us is simply to act as the birthing channel for a form of intelligence that will effectively render humans as we know them obsolete. Other objections stem from a conviction that machines will never capture the spiritual and creative characteristics that distinguish Desmond Tutu from a Toyota Prius, that there is something intrinsic to biology, an animal-biological *essence* that silicon and steel will never possess—simply because they're made of the wrong stuff. The British physicist and überbrain Roger Penrose is arguably the best respected exponent of this view, having articulated it in some depth in two books (*The Emperor's New Mind* and *Shadows of the Mind*). His argument centers around the assertion that "there must be a non-computational ingredient in human conscious thinking"—an ingredient that he suggests may be found deep down in "quantum effects" within the atom.[12] Elsewhere in the mix of objections is the view that any talk of the imminent invention of human-level machine intelligence is largely fanciful anyway, that robotics and AI are still a very long way off doing anything that will rival the human mind. Finally, some thinkers have suggested that we may not be smart enough as a species to work out how we're so smart. Douglas Hofstadter (a longtime commentator on the nature of consciousness) has written, "It could be simply an accident of fate that our brains are too weak to understand themselves. Think of the lowly giraffe, for instance, whose brain is obviously far below the level required for self-understanding—yet it is remarkably similar to our brains."[13] In summary, we don't want it, it's not natural, and it's probably impossible anyway.

One thing is certain. If a conscious humanlike intelligence *is* "computable" (i.e., Penrose is wrong), the processing power to compute it will be within reach of the desktop very soon. Hans Moravec wondered "what processing rate would be necessary to yield performance on par with the human brain?" and came up with the gargantuan figure of one hundred trillion instructions per second,[14] which is one of those numbers that generally makes most of us go "hmmm, I think I'll make a cup of tea now."* To put this number in context, as I was ushered into the world in the early seventies, IBM introduced a computer that could perform one million instructions per second. This is *one millionth* of Moravec's figure. By the dawn of the millennium, chip-maker AMD was selling a microprocessor over *three and half thousand* times quicker (testament to a technological journey that had been populated with continual exponential leaps in processing power throughout the intervening period). This yielded a chip that is still 280 times less powerful than the brain's computational prowess (by Moravec's reckoning) but is a staggering upswing in power nonetheless. As I write this in 2010, Intel has released its "Core

---

*A hundred trillion looks like this: 100,000,000,000,000; and Moravec reckons our brains do this number of discrete calculations *a second*. As Joey said in *Friends*, pointing to his head, "It's not just a hat rack."

i7 Extreme" chip, which is *forty times* faster than the AMD device from 2000 and computes at the mind-numbing speed of 147,600,000,000 instructions per second—or about *one-seventh* of Moravec's figure. At this rate, your new laptop will achieve the same computational speed as the human brain before the decade is out. Soon after that, if the exponential trend continues, your laptop (or whatever replaces it) will have more hard-processing muscle than *all human brains put together*. This will happen sometime around the middle of the century, according to Ray Kurzweil.

Supercomputers have passed Moravec's milestone and it's therefore no surprise to find various projects using them to try to simulate parts of animal and human brains, merging neuroscience and computer science in an attempt to get to the bottom of what's really going on in that skull of yours. Henry Markram, leader of the Blue Brain project (which works by simulating individual brain cells on different processors and then linking them together) believes "It is not impossible to build a brain, and we can do it in ten years." He's even joked (or not, depending on how seriously you take the claim) he'll bring the result to talk at conferences.[15] At the 2009 Science Beyond Fiction event in Prague, Markram told his audience that he believes "the ethereal 'emergent properties' that characterize human thought will, step by step, make themselves apparent."[16] In other words, by simulating the brain at finer and finer levels of granularity, his project will finally allow us to perceive what makes consciousness work from a clear, detached standpoint—it will reveal the laws of intelligence and cognition and we will be able to use them as readily as the laws of motion or computation today. It's no surprise to find that lots of people disagree with him.

Suddenly I'm seeing a link back to my talk with George Church. One way of looking at the human being (and therefore the human brain) is "simply" as a collection of unthinking tiny biomachines computing away—reading genetic code and spewing out "computed" proteins and the rest. We're machines too, just "wet" biological ones. Brooks makes this argument as well in his book *Robot: The Future of Flesh and Machines*: "The body, this mass of biomolecules, is a machine that acts according to a set of specified rules," he writes.

> Needless to say, many people bristle at the use of the word "machine." They will accept some description of themselves as collections of components that are governed by rules of interaction, and with no component beyond what can be understood with mathematics, physics and chemistry. But that to me is the essence of what a machine is, and I have chosen to use that word to perhaps brutalize the reader a little.

In short, intelligence and consciousness are computable, because you and I are computing it right now. I compute, therefore I am. George Church was less "brutal" in his take on the "human machine." "I think of us more and more as *mechanisms*," he told me. "We're starting to see more and more of the mechanism exposed and it just makes it *more* impressive to me, not less. If someone showed me a really intricate clock or computer that had emotions and self-awareness and spirituality and so forth I'd be very, very impressed, and I think that's where we are heading, where we can be impressed by the mechanism."

But something's not sitting right with me, and it's not that I don't like being called a "machine" (believe me, that's nothing compared to some of the heckles I've had). In fact, the machine metaphor kind of makes sense given what I found out at Harvard. I can't quite place my finger on it so I decide to ask Cynthia her view on throwing processing power at a simulation of the human brain as a way to create an artificial intelligence. She does a tiny little snort.

"The bottom line is there's still a long way to go before we can have that simulation actually *do anything*. I mean they can run the simulation but what is it doing that can be seen as being intelligent? How does that grind out into *real* behavior, where you show it something and have it respond to it? I still think there's a lot of *understanding* that needs to be done. I think we're making fantastic strides but I think," she drops to a conspiratorial whisper, smiling, "there's a lot we still don't know!"

Cynthia has nailed the root of my discomfort. Someone can give you the best calculator in the shop, but if you've never learned any mathematics, it's largely useless to you. If the brain is computable, it's not that we won't have the processing power to re-create its mechanisms, but that we're still a long way off working out how to *drive* that simulation. If you'd never learned to read, your eyes could take in the shape of every letter on this page, but it would mean nothing to you, and photocopying it a hundred times (or even inventing the photocopier in order to do so) wouldn't help you either. Just as you had to learn to read, AI and neuroscience research collectively have to tease out not only *what* it is they're looking at, but *what it means*.

After all, there are four hundred miles of blood vessels and a hundred billion nerve cells in that jellylike mass of fat and protein that's inside your head (and that's approximately the same as the number of stars in the galaxy). Trying to understand the interplay of all that cognitive wetware is a mammoth task, and one into which the Blue Brain will no doubt make useful inroads. Cynthia is clear that the project is immensely useful; she just doesn't think the resulting simulation will be conscious. For instance, it will help us ask specific questions about brain chemistry and physiology. This is a notoriously intractable problem because asking questions about how the brain works is a bit like asking my mum whether she enjoyed her dinner.

"Well, I had fish, which reminds me that there was a great deal on fish at Sainsbury's this Saturday, which I found out from talking to Beryl, you remember Beryl? We met her on holiday in Greece and it turned out she lived just down the road in Dunchurch where, by the way, the statue in the square was hit by a car, it was in the paper, front page, did you know your brother's bought a new car . . . ?" (My mum does an amazing thing. She will eventually tell you if she enjoyed her dinner and in the process of getting there will tie up any loose ends she's left hanging during her tangential asides. It all comes together like the video of an explosion being played in reverse.)

Because the physiology of the brain is not unlike my mum's method of answering a question (everything is related to everything else), isolating useful lines of inquiry is quite hard—and it's very likely that one line of inquiry simply doesn't make sense unless it's viewed in the context of ten more. (Imagine watching the playback of the Super Bowl with all but one of the players digitally removed from the action. It'd just seem completely random.) Still, there is hope that some of that interplay will be revealed by the Blue Brain. At the Science Beyond Fiction event, Markram told his audience, "It starts to learn things and starts to remember things. We can actually see when it retrieves a memory, and where it retrieved it from because we can trace back every activity of every molecule, every cell, every connection and see how the memory was formed." These sorts of results are impressive, but returning to the analogy of the Super Bowl, if an alien could analyze a game with a similar degree of scrutiny, describing every move, catch, time-out, even the heart rate of every player at every moment, would they suddenly understand American football? I suspect the answer is "not immediately" but they might have a better idea about the *next* question to ask and the things to look out for and measure in order to answer that next question.

Sure, there's exponential growth in processing power, but the jury is out as to whether there is an equivalent growth in understanding how to use that power more "intelligently" to create (to paraphrase one of Henry Markram's analogies) a concerto of the mind by playing the grand piano of the brain. If there had been this growth, maybe your new laptop would be one-seventh as smart as you are. But it isn't. This is where the strength of projects like the Blue Brain (and Cynthia's work) really lies— as tools to slowly help us pose the *right questions* that will lead to a better understanding of intelligence, emotion and consciousness.

"The thing that's useful about trying to build sociable robots is it pushes us to ask those questions," says Cynthia. "What is life? What is mind? Robots push us to be more precise. They push us to really think hard about what is it that we really mean by those terms." The answers, she believes, are "decades off." In fact, she's very careful to dampen expectations about her own research. I ask her, "What's the longest interaction you've had with a robot?"

"Where it was actually meaningful?"

She pauses for a long time. "It's in the order of like . . ." She hesitates. ". . . tens of minutes, you know? It's short."

Leo's never getting off that box.

It seems to me that the powerhouse that drives and defines real intelligence is *curiosity*. It's why Cynthia's at MIT, why I'm in her office, why you're reading this book, and why a man called Hod Lipson at Cornell University may have just done something profound. Because he's built a machine that seems to exhibit genuine curiosity—a machine that is already working with its creators to generate new knowledge. Accordingly I make Hod my next port of call.

Before I get to Cornell, though, I have two tasks. One is a short gig at the friendly Mottley's Comedy Club in Chatham Street, and the other is to come up with a new word for "robot." I'd asked Cynthia if she thought the word was restrictive, given all its associations in popular culture. Did it maybe fail to get across her idea of a wide range of sociable machines with their own take on emotions? "Yes, I'd like to come up with something different," she'd said, and then looked me straight in the eye. "How about *you* come up with a better term? And I'll start using it!"

The word I come up with on my journey upstate is "AInimal" (pronounced either "ay-nimal" or "ay-eye-nimal" depending on your preference). A creature with an artificial mind, but a creature nonetheless.

5

"Like Explaining Shakespeare to a Dog"

I have no special talents. I am only passionately curious.

—*Albert Einstein*

How do you find a truth?

Here's one. Some students at Cornell University smoke weed. How do I know this? Because shortly after arriving in the small town of Ithaca I find myself standing outside Insomnia Cookies. "Warm Cookies Delivered Late Night" says the purple neon sign in the window. I don't smoke "hippie lettuce," but I've been around enough connoisseurs of the herb to know that they tend to get a craving for baked snacks ("the munchies"). This makes putting a late-night warm cookie store (that will deliver!) on a university campus nothing short of *retail genius*. (I subsequently discover that Insomnia Cookies has outlets on eighteen campuses throughout the States.)

It is this kind of creative thinking that AInimals find hard. It's also hard for them to infer a wider truth from limited data. There isn't an AI in the world that has spent one night too many listening to stoned, cookie-eating students spout cod-psychology to a soundtrack of space rock—and so there's not one that could, given sight of a single cookie store, make the deduction I've been able to: late-night campus cookie store = "the munchies" = stoned engineering students.

That said, sitting just a few blocks from where I'm standing is a computer brain that *can* find truth from looking at slivers of data, using previous experience to make its deductions. It's a brain born of the body now doing some serious intellectual work—a "sense and action" robot that has graduated.

I find one of the machine's creators, Professor Hod Lipson, in his lab on the second floor of Upson Hall, a nondescript university building just off the "Engineering Quad" of Cornell's verdant campus. He has a round, solid but not chunky frame, thick curly black hair, and he looks healthy and happy. His features are open and innocent but strong and wide, giving him a big smile and a certain physical power too. If Tom Hanks had become a powerlifter instead of an actor, he'd probably look a bit like Lipson. There's no *swagger* about Hod Lipson, but there is evident self-belief. You get the feeling Hod knows exactly what he wants to achieve, but that he also intends to have fun doing it. And while most of the scientists I've met are driven by an almost insatiable curiosity, Lipson takes curiosity to a new level, literally. He's curious about curiosity.

"Artificial intelligence is a moving target," he says, taking a seat opposite a table piled high with AI textbooks and professorial paperwork. "I want to create something where nobody can argue it *isn't* intelligent. I was thinking, what's an unmistakable, unequivocal hallmark of intelligence? I think it's creativity, and particularly *curiosity*."* This is why I've come to see Hod. If our future is to include truly intelligent machines, then they'll need to be curious and creative ones. If we can produce such AInimals, they could radically reshape how we go about finding and applying new knowledge, heralding a revolution in *thinking* similar to the one commercial robots have already achieved in manufacture.

---

*It's a popular view. In his paper "Celebrating 75 Years of AI," Jürgen Schmidhuber wrote, "No theory of AI will be convincing if it does not explain curiosity and creativity."

I get straight to the point. "Does a curious and creative machine mean a sentient machine?"

"Well, what does that mean?" asks Hod. "I have to push you on what you mean by 'sentient.'"

Bollocks. Less than a minute in and I'm getting strong *déjà vu*. Cynthia tried to sidestep the question of what constitutes machine consciousness. Hod wants *me* to answer it, which seems a bit unfair given he's the professor. I do the only thing I can.

"Well, let me ask you," I say. "What do *you* mean by it?"

Hod pauses. I'm not sure he was expecting a return serve, especially one that in any decent rule book would be considered cheating.

"I interpret it as deliberate versus reactive. Er . . . humanlike . . ." He pauses. "I don't know."

I claim a draw.

Lipson's lab became famous (in robotic circles) for its work building AInimals that are arguably *self-aware*. The lab's Starfish robot built by Hod, Josh Bongard and Victor Zykov is notable for learning to walk from first principles.* It wasn't programmed to walk; instead, Lipson and his team programmed it to learn about itself. It then used *this* knowledge to work out how to move. It's another example of Brooks's "sense and action" model, and in understanding this mechanical starfish I start to see what Brooks was getting at when he wrote, "simple things to do with perception and mobility [are] a *necessary basis* for higher-level intellect."[1]

One of the characteristics of a "higher-level intellect" is a *self-image*—an understanding of where you begin and end, what's the world and what's you. It's something babies develop during the first years of life.[2] Creating this physical self-image is crucial in understanding the relationship between your foot and the floor (or indeed your hand and

---

*Just like Leo, the Starfish is sitting inert and almost forgotten on a shelf in the lab when I visit.

the fire) and babies learn these things through physical experimentation. Child psychologist Dr. Richard Woolfson has written, "Your baby is a bit like a scientist, and her self-awareness grows as she explores."[3] Just like a baby, the Starfish developed its self-image through physical experimentation.

"So it moved its motors, it sensed its motion and then it created various models of what it might look like—'maybe I'm a snake? maybe I'm a spider?' We told it to create multiple different explanations that might explain what it knows so far." (In this instance, "model" is science-speak for "mental image.") The robot then tested each model by sending them into competition with each other.

"We tell it to create a new experiment that creates disagreement between the predictions," says Lipson.

One of the key strengths of *human* intelligence is its ability to contradict itself. Our capacity to hold conflicting ideas allows us to argue things out in our heads and collectively as a society. What Hod and his team did was build this productive contradiction into the Starfish.

"We set off *two* lines of inquiry. So one of them is the thing that creates models and the other is the thing that asks questions. They have a predator-stalks-prey relationship. The questions basically try to break the models." Over time the predictions of the competing models get closer and closer to what actually happens until one model is deemed sufficient for the robot to say, "This is what I look like."

It has to be said that the Starfish isn't the most graceful creature in the world. It doesn't so much walk as stagger and flop forward. Less Ginger Rogers, more gin and tonic. Still, the achievement is not to be sniffed at. It had no parents and no role models. This was an AInimal actively learning to do something no one had taught it to. But what happened next is why I've come to Cornell. The team's AInimal brain, born of physicality, started to do some extraordinary things in the symbolic arena (you know, all that "educated male scientists" stuff). It started to churn out scientific truths.

One could argue the search for truth is history's guiding force. There have been libraries of books written (and lost) on the subject. But if someone said to me, "Go on then, history of truth in five minutes," I'd probably reach for two key figures—Socrates (born Greece, 469 BCE) and Francis Bacon (born England, 1561), not least because they both died in interesting ways (which, while not great for them, is useful for storytelling).

Socrates was put to death by the state of Athens for "refusing to recognize the gods recognized by the state" and "corrupting the youth" (explaining perhaps why Black Sabbath rarely toured in Greece).[4] Despite the opportunity to escape his fate, Socrates placidly took poison hemlock prepared by the authorities. Francis Bacon, many believe, died as a result of trying to freeze a chicken. It might seem odd, therefore, that both are seen as key figures in the history of *reason*.

Both Socrates and Bacon were very good at asking useful questions. In fact, Socrates is largely credited with coming up with a *way* of asking questions, "the Socratic method," which itself is at the core of the "scientific method," popularized by Bacon during the Enlightenment—a period of European history when "evidence" and "faith" had an almighty bun fight and the balance of power between church, state and citizen was questioned. Lots of philosophers and scientists challenged the prevailing orthodoxy of religious authority by saying, "We need to make decisions based on evidence and reasoned debate, not on sacred texts and religious faith," to which the church replied, "Yes, but we own most of the land, plus people really *like* God. Ask them."

The Socratic method disproves arguments by finding exceptions to them, and can therefore lead your opponent to a point where they admit something that contradicts their original position. It's powerful because it can get people to admit to themselves that they're wrong. It's pretty good at exposing your own (as well as others') prejudices and gaps in reasoning.

Lawyers use it a lot. (Don't let this color your opinion of it. Lawyers also use toilet paper and you're not about to reject that idea.)

Here's an example of it in action: The period of the mid-eighties in the Stevenson household was accompanied by a soundtrack of hard and "progressive" rock coming from my elder brothers' bedrooms. My dad, in retaliation, would argue that "people only play electric guitars because they can't play real ones" (by which he meant acoustic guitars played by nice chaps called Julian with sensible haircuts, as opposed to electric guitars played by long-haired geezers called Dave and Jimmy).

First step in Socratic questioning: assume your opponent's statement is false and find an example to illustrate this—Pink Floyd's David Gilmour playing "Wish You Were Here," for instance. Clearly Gilmour can play a "real" guitar as well as an electric one and Dad had to grudgingly accept the fact. At this point he would assert that Dave Gilmour was "the exception that proved the rule."

Step two: if this has not convinced your opponent of their error, take their original statement and restate it to fit their new position. "So, Dad, you're saying that people only play electric guitars because they can't play acoustic ones, *except* for Dave Gilmour, who can do both?" Then return to step one (finding clear examples that contradict this new statement).

Ironically this led to my brothers asking my dad to listen to far more Black Sabbath, Pink Floyd, Aerosmith and Led Zeppelin than if he'd kept his theory to himself. In painfully slow increments my dad would accept that, yes, Tony Iommi, Joe Perry, Jimmy Page, Steve Howe, and, okay, all right! Gary Moore, Ritchie Blackmore, Eddie Van Halen and a whole host of other rock fiends could indeed play "real" guitars. Eventually he had to admit *the truth*—which was not that the rock musicians my brothers listened to weren't talented, but that he *just didn't like rock music*. This example may seem trivial (especially if you don't know my dad and/or aren't a fan of "classic" rock) but across the board, Socratic questioning can expose fundamental new insights, which is why it's long been of interest to those trying to get robot brains to learn new things. That said, it has its dangers. Despite being undoubtedly one of the most useful thinking tools ever devised, it can also annoy the tits off people—as the man who lends it his name found out to his cost. The story is that Socrates used his technique to prove a lot of bigwigs in Athenian society were wrongheaded—and so they killed him.

By the time Francis Bacon went to university, teachings of one of Socrates' students, Aristotle, had become the accepted way to conduct "scientific inquiry." Aristotle had pioneered "deductive reasoning," the practice of deriving new knowledge from foundational truths, or "axioms." It was generally believed that if you got enough deep thinkers together to have a solid debate, scientific truth would be teased out over time.

This have-a-long-and-involved-chat method worked well for mathematics and still does, because the axioms are long established (e.g., the basic "mathematical operations"—plus, minus, divide, multiply) but it wasn't as useful for finding out new knowledge relating to the physical world. Much to Francis's dismay it seemed that "science" involved sitting around in armchairs and talking. Nobody was getting off their arse and observing anything new or doing any experiments. Nobody was discovering the "Axioms of Reality" (arguably a good name for the sort of progressive rock band that would have my dad shouting "Turn that racket down!"). Just like early robots these scientists weren't really *in* the world (something that still afflicts many senior academics today).*,5

In common with Socrates, Bacon stressed it was just as important to disprove a theory as it was to prove one—and real-world observation and experimentation were key to achieving both aims. Bacon also saw science as a collaborative affair, with scientists working together, challenging each other. All these remain hallmarks of good scientific practice today—observe, theorize, experiment . . . and then try to prove yourself wrong—all in collaboration with peers who can (and should) give you a hard time. It's important to note that Bacon himself wasn't a

---

*As "creative learning" advocate Ken Robinson remarks, "There's something curious about professors in my experience. . . . They're disembodied . . . they look upon their body as a form of transport for their heads. . . . It's a way of getting their heads to meetings. If you want real evidence of out-of-body experiences get yourself along to a residential conference of senior academics and pop into the discotheque on the final night. And there you will see it; grown men and women writhing uncontrollably. Off the beat."

distinguished scientist; his main contribution was to describe and champion an evidence-based scientific method. That said, he did do the odd experiment, including the one that killed him.

While traveling from London to Highgate with the king's personal physician, Bacon wondered whether ice might be used to preserve meat. The two men got off their coach, purchased a chicken and stuffed it with snow to test the theory. In his last letter, Bacon is said to have written, "As for the experiment itself, it succeeded excellently well" but the act of stuffing the chicken led to his contracting fatal pneumonia, which resulted in his death a few days later.[6] This is possibly the only instance of Bacon being killed by eggs.

Back in modern-day Ithaca, the spirits of Socrates and Bacon were coming out of a robot.

Hod (along with colleague Michael Schmidt) wondered if the "sense and action" brain from the Starfish could go beyond working out what its own body looked like and begin to reach useful conclusions about the wider world. To put it another way, having gone through the process of building an accurate mental model of *itself*, could it make accurate mental models of other things?

Their first idea was to give the robot brain the ability to set up the starting position of a "double pendulum" before letting it fall. It was then fed the results of each experiment—which had been recorded using motion-capture technology—allowing it to accurately assess the pendulum's motion.

A double pendulum is basically one pendulum, with another pendulum stuck to the bottom of it—two sticks joined end to end by a free-moving hinge. While the top half swings from left to right, the bottom half soon starts to go a little crazy. Because the bottom section isn't attached to a stationary point (like the top pendulum) but to something that's already in motion (the lower end of that swinging top pendulum), it will swing left, swing right, spin round clockwise, or counterclockwise, seemingly at random. Lipson and Schmidt chose the double pendulum because it's a good example of a system that is simple to set up but quickly exhibits chaotic behavior—and would therefore be a good test of the robot brain's skills. The results were startling. In fact, the Starfish brain went a long way to deriving the laws of motion, previously sweated over for decades by the likes of Isaac Newton.

And it did it in three hours.

It followed the same process as when it sat in the Starfish—guessing at models (mathematical equations) that might explain what it had seen so far, then creating new experiments (new starting positions for the pendulum) that targeted the areas of most disagreement between equations it had guessed at so far. "With the double pendulum it very quickly puts it up exactly upright, because some models say it's going to fall left and some models say it's going to fall right. There's disagreement. It's not a passive algorithm that sits back, watching," says Hod, smiling. "It *asks questions*. That's curiosity." One of the key things this AInimal was curious about were "invariants."

Imagine your lover is having a hard time at work, putting in long hours to meet a crucial deadline. They feel guilty this takes them away from you, so to make up for giving you less attention they might bring you a gift. It's a way of showing you affection even when their time is limited. You're getting the *same amount* of love, it's just *distributed* differently. (This is known as Stevenson's Second Law of Relationships.*) If you're seeing a mathematician, they might explain this by saying, "Darling, my love for you is invariant!" (And you may want to consider dating a poet.)

"Let's say you have two sets of measurements coming in from some experiment and they wiggle up and down in some complicated way," says Hod. "Our machine tries to find out what's *constant* about them. Maybe if you add the two figures together, you always get the same number? Maybe they're always in the same ratio, even though the numbers shift? Maybe it's more complex—perhaps the sum of their square roots is the thing that stays the same. You wouldn't immediately see that just from looking at the numbers going up and down."

---

*Stevenson's First Law is *never* date your landlady.

It's these hidden "invariants" that often provide crucial insights into what's guiding a system's behavior. "Asking what stays constant about the pendulum can help you discover that the force you apply to something is always equal to its mass multiplied by its acceleration; Newton's Second Law of Motion."

Just like humans, machines seem to learn best when they ask their own questions and find their own answers, rather than being given huge amounts of data to absorb. "In the real world you can't sit back and just watch as the data comes in," says Hod. "You have to probe, because collecting lots of data takes time and can be risky." By contrast, Lipson and Schmidt's machine brain "only ever sees what it asks for. It does not see all the data." In fact, Lipson decided to compare the efficiency of this "active" method of inquiry against a more traditional passive "here's all the data, what can you tell me?" method. "It doesn't work. It has to go through *reasoning*."

Speaking with Hod, I see a certain hemlock-taking philosopher and a chicken-freezing experimentalist partially reincarnated in machine form. The robot's programming consigns inaccurate models to the dustbin by getting it to admit that others offer a better explanation of the real world (hello, Socrates) with evidence won via experimentation (hello, Bacon), sniffing out fundamental, unshifting constants hidden in the data. Lipson has created a computer-based way of asking good questions—one of the hallmarks of true intelligence.

"Physicists like Newton and Kepler could have used a computer running this algorithm to figure out the laws that explain a falling apple or the motion of the planets with just a few hours of computation," said Schmidt in an interview with the U.S. National Science Foundation.[7]

However, we're still a long way off what Hod (or anyone else) would call a truly intelligent machine. It still takes a human to work out if anything the machine has found is useful. The machine didn't *know* it had found laws of motion; it took Hod and his colleagues to recognize the equations. "A human still needs to give words and interpretation to laws found by the computer," says Schmidt. Again, the sticking point in AI is assigning *meaning* to something.

One of the areas in which Hod's "brains" could turn out to be useful is in cracking problems where there is lots of data, but we haven't much idea what's going on. In fact, as soon as Hod published his results, numerous people with acres of data were beating a path to his door, including heavyweight data generators like the Large Hadron Collider at CERN. "The people at CERN said, 'There is this gap in a prediction of particle energy. Here's data for three thousand particles. Can you predict something?'" The result was a strange mix of elation and disappointment. "We let it run and it came up with a beautiful formula," says Hod. "We were very excited but it was a famous formula they already knew. So for them it was a disappointment. . . . But for us . . . we rediscovered something that people are famous for."

Again, the crucial insight came from humans. Without their finding meaning, the results are largely worthless (which is not to say the time saved is not incredibly useful). I'm reminded of a scene in Douglas Adams's *The Hitchhiker's Guide to the Galaxy* in which a hypercomputer, Deep Thought, is built by a race of supersmart humanoids to compute the answer to the ultimate question of "life, the universe and everything," which after millions of years of deliberation it announces to be 42, before suggesting that the humanoids don't really know what "The Question" is.

No one understands the irony in this story more than Hod Lipson. "In biology there are many systems where we do not know their dynamics or the rules that they obey." So he set his machine looking at a process within soil bacterium. True to form, the program generated an equation in double-quick time. But what did it mean?

"We're still looking at it," says Hod with a smile. "We're staring at it very intently. But we still don't have an explanation. And we can't publish until we know what it is."

"You don't understand what it's saying?"

"No," says Hod.

"It's your 42 moment?"

"Yeah," says Professor Lipson, smiling ruefully.

Several months after my visit I e-mail Hod to see if he's gotten any-where with the mystery equation. "We're still struggling," he writes. "We've been trying for months to get the AI to explain it to us through analogy. But we don't get it. Maybe it's hopeless. Like explaining Shake-speare to a dog," suggests Hod, echoing Douglas Hofstadter's idea that there might be things we're just not built smart enough to understand. Even still, the machine's achievement is important.

"In science you go from observations which produce data, to models which produce predictions, to underlying laws," explains Hod, "and from there you go to *meaning*." Hod's robotic brain can help slash the time those steps take. "We can go from data *straight to laws*, whereas previously people could only go from data to predictions. So now a scientist can throw it some data, go and have a cup of coffee, come back and see fif-teen different models that might explain what is going on. That saves a lot of time. Previously coming up with a predictive model could *take a whole career*. Now at least you can automate that so you can focus on meaning."* That's a powerful enabling technology, because it leaves more time to think. More time to do that thing that humans are good at and comput-ers aren't—attributing meaning to things. Hod is doing for thinking what dishwashers have done for after-dinner conversation.

I ask him if his brain could come up with a model of how to learn. After all, if it can come up with a model of itself, or a model teased out of

---

*This isn't the only example of a robot helping to automate the process of science. In 2009, Aberys-twyth University in Wales announced that its "robot scientist" *Adam* had discovered the role of twelve different genes in yeast cells. By removing individual genes from different cell strains and then observing them as they grew, it deduced which gene did what. "It's like a car," said Professor Ross King, who led the development of Adam. "If you remove one component from the engine, then drive the car to see how it performs, you can find out what that particular component does." The specific role of these genes had evaded human researchers for decades, but Adam's ability to churn through thousands of repetitive experiments (with a human on hand to provide the experiments' feedstocks and take away waste products) without tiredness or boredom was the key to solving the problem.

data sent from CERN, or a model of energy conservation derived from looking at a pendulum, why not get it to observe something learning and see if there are insights that could be applied to machine learning and the further development of AI?

Hod laughs. "That's what we're working on now. We're working on what we call 'self-reflective systems' where they are thinking about thinking." Hod has clear reasons for doing this. "If you want to get to human-like intelligence, you need a brain that can think about thinking. . . ."

Hod is taking his model of "coevolutionary AI" to the next level. Instead of modeling robot physiology and learning about walking, the motion of pendulums or energy signatures of subatomic particles, he has one brain trying to model how another one learns and then, in true Lipson style, he's asking one to challenge the other in order to find out more. In this way, one brain builds a model of how the other learns, and can start to make helpful suggestions.

"That's self-reflection," says Hod. He adds, "That's important in life. You can learn things the hard way, or you can think about how you've been thinking."

Curiosity? Self-reflection? Social interaction? A theory of mind? Thinking about thinking? While AI still has a long way to go, the questions researchers like Hod and Cynthia are raising (and the results they are getting) show that the field is far from stagnant. It's clear that machine intelligence (both "social" and "symbolic") is advancing, if in very small increments. The question is, how might this impact you and me? How might it all play out?

The first option is that AI and robotics continue to make advances but are unable to achieve a humanlike or human-level intelligence. Penrose is right and consciousness is "noncomputable," or Hofstadter got it right when he said that, even if it is computable, we may not be smart enough to work it out. (The computer scientist Alan Perlis once remarked, "A year spent in artificial intelligence is enough to make one believe in God,"[8] infer-ring that a failure to create a conscious machine will lend more weight to the argument that we all have an immortal soul.) In this scenario, machines like

Hod's will keep churning out models and equations, but it'll still be humans making (or failing to make) sense of them. The rate at which knowledge advances will go up a good few notches, aided by ever more powerful technology, but the ultimate speed limit will be how quickly we can take meaning from what our machines are helping us discover.* Cynthia's robots will get ever more social, and her research will help us interface with our technology in ways more natural to us, but this sociability will be a result of some impressive computer processing that falls short of machine consciousness. As legendary AI researcher Marvin Minsky remarked upon seeing Leo, "My objection to Leonardo is, it's just a trick. It doesn't really have emotions. It just knows how to fool you into thinking it does."[9] If Minsky is right, it still doesn't take away from Cynthia's key achievement to date: that we can sometimes *accept* her robots as emotional and sentient, even if they're not.†

Option two is where machines *do* achieve a level of intelligence we accept as undeniably sentient, something that goes beyond trickery and enters true personhood. Somewhere on this road, we'll have to deal with the sticky issue of whether robots have, er, human rights. As Nick Bostrom, whom I met in Oxford, points out, "Whether somebody is implemented on silicon or biological tissue, if it does not affect functionality or consciousness, is of no moral significance. Carbon-chauvinism is objectionable on the same grounds as racism."[10] People complain about robots taking over their jobs now. Imagine the reaction when a robot can sue for unfair dismissal on the basis of "machinism."

---

*Shortly after our meeting, Hod puts the code for his brain online, giving it the name Eureqa (pronounced "eureka") for anyone to feed data into.

†Cynthia tells me Minsky is "missing the point" by "only focusing on the internal aspects of emotion and not the interpersonal and using human emotion as the gold standard. Dog owners don't say their dog's emotions are a 'trick' or 'fake' even though they are not human emotions." Indeed. On a return trip to Boston to see Ray Kurzweil, I pop back into the Personal Robots Group and see its latest sociable robot, Nexi, in action. In tune with the lab's focus on human-robot relationships, Nexi is interacting with a young boy no older than eight who finds her humanlike tracking of his movements as he dances in front of her enthralling. Despite being made of molded white plastic, Nexi's face expresses a whole gamut of emotions—her big eyes blinking, her white plastic "eyebrows" moving, her mouth expressing slack-jawed boredom (when not much is happening) to tight-lipped interest (if the boy is doing something intriguing) or annoyance (if the boy gets too close). It's startling to see how quickly both the boy and those of us watching accept Nexi as somehow sentient and alive.

If the AInimals do get truly smart, we might enjoy a confederacy of man and machine. We remain separate but work together. This is a world where sometimes our machines "explain things to us," says Hod. Some machines are good (C-3PO) and some machines are questionable (HAL 9000). Machines become part of humanity. They're "us" as much as the French and English are "us" to each other—foreign, sometimes hard to get along with, at other times the people we'd prefer to spend our holidays with but still, in the final analysis, part of our global society.

But there are, of course, more apocalyptic visions of this future. After the machines get smart, they quickly engineer themselves to get smarter still. After all, we'll have worked out how to "compute" intelligence and the AIminals, benefiting from the continuing spiral in processing power, will compute that intelligence faster than we can. Within a few short technology generations, they could leave us behind in a cognitive backwater called "humanity." We become to our AInimal progeny what primates are to us. At this point we hope they feel some responsibility for their ancestors and don't break us down for raw materials, try to hunt us to extinction (*Terminator*) or (yes, *Matrix* fans) use us for batteries. They rule us, and our fate will be in their hands.

Option three is that man and machine merge. As I found out during my chat with Nick Bostrom, this is already happening. In the opening pages of *Robot* Rodney Brooks writes, "Recently, I was confronted with a researcher in our lab, a double-leg amputee, stepping off the elevator that I was waiting for. From the knees up he was all human, from the knees down he was robot." In late 2009, Robin af Ekenstam became the first recipient of "SmartHand"—a prosthetic mechanical hand that not only replicates the movement of its human equivalent, but also provides Robin with a sense of touch (the replacement hand is wired to existing nerve endings in the stump left after he lost his original hand to cancer). "I am using muscles which I haven't used for years. I grab something hard, and then I can feel it in the fingertips, which is strange, as I don't have them anymore," said Robin. "It's amazing."[11]

It's not a short leap of the imagination from mechanical body parts to a world in which we also merge with mechanical brains. This future comes in two flavors. The "vanilla" version is a bio-interface that allows us to access technology that gives us extra capacities, a bit like plugging an external hard drive into your laptop. You won't need to forget anything because you'll have petabytes of hard drive storage wired into your brain that you can recall data from at will. Difficult sum? Outsource it to your math chip. In this scenario the consciousness is all biological, but it's got a load of new toys to play with.

The other flavor of mind/machine merge is where we unite our human consciousness with machines, running part or all of our brains on this new platform. When I return to Boston, Ray Kurzweil will tell me we'll use technology to amplify and focus our intelligence and conscious-ness, replacing our brains with something more powerful, perhaps mov-ing away from biology altogether. "We won't be limited to a neo-cortex that fits into a less than one-cubic-foot skull, and we certainly won't run it on a chemical substrate* that sends information at a few hundred feet per second, which is a million times slower than electronics," says Ray. "We can take those principles and reengineer them and we're going to merge them with our own brains."

This makes Nick Bostrom's argument that "whether somebody is implemented on silicon or biological tissue . . . is of no moral signifi-cance" takes on a personal resonance, and brings a whole new meaning to the phrase "a meeting of minds." The AInimals don't leave us behind or exterminate us, because we *are* the AInimals.

"We will become a merger between flesh and machines," writes Rodney Brooks in *Robot*. "We will have the best that machineness has to offer, but we will also have our bioheritage to augment whatever level of machine technology we have so far developed. So we (the robot-people) will step ahead of them (the pure robots). We won't have to worry about them taking over." In a 1995 paper entitled "Will Robots Inherit the Earth?" Marvin Minsky concluded, "Yes, but they will be our children."[12]

---

*In this instance, what Ray means by "chemical substrate" is your current human brain.

This makes many people feel uncomfortable. Most of us can make peace with some physical encroachment on our bodies (I'm very pleased with my contact lenses for example). I'm glad Brooks's student can walk. And I think most of us are probably okay with brain interfaces that help the formerly disabled use new prosthetics. I'm truly heartened that Robin af Ekenstam can now eat a meal with a knife and fork using a robot hand controlled directly by his thoughts. Who would deny him the opportunity to regain his dexterity and sense of touch?

But merging our brains wholesale with technology? Or simply doing away with our neurons altogether and "porting" ourselves completely over to silicon? I have to admit that writing this book would have been a whole lot easier if I'd had the Internet and the British Library wired into my head. And in a brain-meets-computer future you could read it as a download in milliseconds. Depending on how you view the world (or perhaps how busy you are), either the idea is strikingly brilliant or rather misses the point. It's the whole "therapy *vs.* enhancement" argument I discovered at the beginning of my journey in Oxford.

It strikes me that one of the problems I have in making sense of all this is that ideas as powerful as intelligent AInimals and man-machine mergers have the ability to instantly grab my imagination and emotions. I am instantly *there* somehow, imagining a dystopian or utopian future depending on my mood, or the tone of the last sci-fi movie I watched. But it's worth remembering that the idea of the intelligent machine has been with us for over two thousand years. (Around 300 BCE, "artificer" Yan Shin reportedly showed a humanlike automaton to King Mu of Chou.[13]) While machines like Hod's and Cynthia's are "smart" in some ways, one thing they've both given me is the clear realization that if machine consciousness is possible, it's not suddenly going to leap up overnight and say, "Hi, I'm here to be your new best friend/turn you into a human slave."

"Intelligence" and "consciousness" are still mysteries, and although the fields of robotics, AI and neuroscience are doing much to unlock their secrets, I've seen nothing that leads me to expect that a robot army (or laundry detail) will be knocking on my door anytime soon. The journey to creating intelligent machines is progressing in baby steps, which means we should have time to negotiate our interactions with them as they slowly get cleverer (a bit like we do with our kids). This isn't to deny the extreme speed with which information technologies advance. Neither does it argue that we won't have the raw computational or manufacturing *power* to make such machines. It's just I've become convinced that the ability of machines to understand *meaning* is moving much slower than the raw technology. It's not the parts, it's the (intellectual) labor.

To me this suggests that if intelligent robots do enter society, they'll do it slowly, gradually getting smarter and more reliable as we (no doubt with their help) unpick the mysteries of cognition. There is no overnight robotic revolution waiting in the wings. Certainly a robot-human society will be *different*, but I can't find a single shred of evidence that automatically points to robots becoming our masters. Other human beings scare me far more than robots do. And if the robots do at some point start to vastly outstrip us intelligence-wise? We'll jump right on that bandwagon and go with them.

As my bus pulls out of Ithaca, George Church's parting words come back to me: "Evolution is far from over. It's accelerating, and something is going to replace *Homo sapiens*. Is it going to be wetware, or is it going to be silicon? Is it going to be genetically engineered, or is it going to be electrically engineered? Or, is it going to be some hybrid?" My bet is on the hybrid, Rodney Brooks's merger of flesh and machines, which we're already seeing in medicine and, I suspect, will slowly continue. It sounds weird now, but the pace at which we'll accept machine encroachments on our bodies and minds (beyond those we already have) will probably come down to one simple principle (and our discomfort with it will melt quickly as a result)—and that is simply that the technology *doesn't suck*.

Not that long ago IVF was considered highly controversial. Indeed, the Catholic Church maintains its objection to the technique on the grounds that in IVF, "the generation of the human person is objectively deprived of its proper perfection: namely, that of being the result and fruit of a conjugal act."[14] (Catholics are instructed that separating sex from procreation is contrary to doctrine.) However (and with no disrespect to Catholics), mainstream opinion has generally gotten over its qualms about "man playing God" when it comes to helping couples with fertility problems conceive. "In vitro fertilization was a big ethical problem, a big policy issue," George Church remarked. "But the instant it worked, and you started having people zipping about with their cute babies, everyone with reproductive issues was writing to their congressmen, and asking their doctor about it. They want that service, because they can see it, and see it can work. The argument flips a hundred and eighty degrees. It became *unethical* to *deny* access to IVF." A brain link to Wikipedia sounds freakish now because it's not been done before. But if and when the technology starts to work seamlessly, then I can guarantee some people will be (unsportingly) taking them to pub quizzes.

It's important to note that no one will *force* you to implant anything. You can reject brain interfaces and night vision (and whatever else) just as you can decide to live without a mobile phone, television, the Internet and cheese. But it's likely that society, rather than rejecting these advances, will begin to insist it's *ethical* for us to have them, echoing the IVF story. You might take some enhancements and not others. There will be consequences to those decisions, but they're yours to make.

So for now the most intelligent machines on the planet are us. And it looks like in the future, they'll be "us" too.

How quickly intelligent robots and man/machine interfaces become commonplace is linked to another technology that has the potential to radically reshape our world and our economics—a technology that even has the potential to spell the end of capitalism. It is perhaps the most powerful technology ever created, and will play a game-changing role in everything from AI, to medicine, to manufacture. It's already beginning to enter our lives, but we're right at the beginning of its journey to ubiquity. Alternatively characterized as the harbinger of an unstoppable techno-rampage that could engulf us all, or a revolution that will reduce the cost of nearly everything, it promises to usher in a transformation in manufacturing more potent and far-reaching than the birth of the industrial age. It is the third pillar of what futurists call the "GNR revolution" (the "G" standing for "genetics" and the "R" representing "robotics").

It is "N"—nanotechnology. And it threatens to change the world in ways that will make your head spin.

"Invisibly Small and Magical"

God dwells in the details.

—*Ludwig Mies van der Rohe*

"Most of them are higher dimensional beings. They don't *need* a spaceship to come here," says the man with long gray hair sitting two tables away. "You have to understand there *already is* a United Federation of Planets," he says (borrowing *Star Trek*'s fictional governing body to illustrate his point). "We're being observed by twenty-six different alien races. Of course, the unethical ones—the Recticulans, the Greys, the Nordics—are ignoring the Prime Directive," he adds matter-of-factly (nicking *Star Trek*'s fictional central dogma too).

There's a piece of Silicon Valley folklore that says if you really want to know what's going on in the technology industry, go and sit in a Hobee's (a small chain of eight restaurants that can be found in low-rent locations throughout the area) and listen to the conversation around you. The story is that you'll learn more this way than by reading any amount of paid-for analysis. I have clearly chosen the wrong seat.

This is California, where childhood *Star Trek* fans, whether they are UFO conspiracy theorists or high-tech multimillionaires, rub shoulders—and where I have traveled to talk, appropriately, about both the fantastical and high tech. But in contrast to the frame-shifting potential of the alien civilizations currently being discussed (very loudly) by the man with the hair (although he's now telling a story about how he resisted arrest over a marijuana charge*), my conversation today will deal with something very real and born of this planet. It may sometimes sound like science fiction, but it'd be a mistake to write it off as such, even though it could change our view of the world just as fundamentally as if ET landed on the lawn of the White House and asked to use the bathroom—and it could radically reshape our future.

I've come to investigate a technology of which the highly regarded future commentator and "eco-pragmatist" Stewart Brand has written: "The science is good, the engineering feasible, the paths of approach are many, the consequences are revolutionary-times-revolutionary, and the schedule is: in our lifetimes."[1] It's a technology that has the potential to end industrial capitalism, revolutionize energy production, boost the power of medicine, deal a deathblow to nearly any resource crisis and ask you to review your relationship with your cleaning habits. Brand is talking about nanotechnology and I'm here to meet K. Eric Drexler, the man largely attributed with inventing it. In 1986 Drexler wrote a book called *Engines of Creation: The Coming Era of Nanotechnology* (now available for free on his website) and pretty much everyone who read it sat up and thought, "Well, *that* changes everything."

But what is nanotechnology? Here's where I hit my first problem. It has come to mean anything outside of biology that we try and do, engineering-wise, on the scale of nanometers—that is, on the scale of one billionth of a meter, or "the ridiculous 'invisibly small and magical' definition," Eric later tells me, smiling. I feel I need something a bit more concrete.

How small is a billionth of a meter? Think small. Very small. Look, I know you think that's small, but, er, go smaller. Sorry, you're still not there. Okay, have you got to the point where it's too small to comprehend? So small you can't really picture it? Then you're still *way too big*.

Think of it this way. The average width of a human hair is just over one-twelfth of a millimeter. You'd have to chop up the 30 cm ruler you

---

*I wonder if it was outside a branch of Insomnia Cookies. . . .

had at school into *3,750* equal slices to get sections the same width as that hair. A nanometer is roughly eighty thousand times *smaller than that*, eighty thousand times smaller than the diameter of a human hair. To chop your ruler into nanometer-sized chunks, you'd have to slice it into *three hundred million* equal pieces. It really is just too small to think about. Except a lot of people *are* thinking about things on this scale, many of them because they read *Engines of Creation* and decided to become one of those newfangled nanotechnologists.

I come to the conclusion that thinking only about scale is distracting me from what Drexler's original vision for nanotechnology is really about—manufacture. Sometime nanotech historian Ashley Shew writes, "Nanotechnology is so interesting because it is not one subject of study. Rather, it is a scale that has brought together researchers, innovators, and engineers from sometimes radically different fields. Disciplines collide on the tiny."[2] We're used to the collision of disciplines in traditional manufacturing. Skills of design, robotics, materials science and electrical engineering come together to make everything from your car to your hair dryer. Those same disciplines and others are colliding again, just on a smaller playing field. This changes some of the rules of the game (how you make stuff) but the game (building the stuff) stays the same.

Focusing on *what can be made* as opposed to classifying things by their size is one reason the Industrial Revolution wasn't called the "yard-long technology revolution" and it's the reason Eric now prefers the phrase "molecular manufacturing" over the simpler "nanotechnology," which has become a catchall for anything really tiny. It's this nano-*manufacture* I've come to California to talk about, because even the most cautious commentators believe it could change everything.

Deep breath then. Here I go again.

The promise of nanotechnology is based on the idea that if we can engineer things with atomic precision, we can develop materials and products with extraordinary properties. Changing the structure of something at the atomic level can cause it to exhibit unusual and often astounding properties. Nanotechnology is to matter what a phone booth is to Superman— it can bring about a transformation that unleashes superpowers.

Drexler opens *Engines of Creation* with an obvious but powerful observation:

> Coal and diamonds, sand and computer chips, cancer and healthy tissue: throughout history, variations in the arrangement of atoms have distinguished the cheap from the cherished, the diseased from the healthy.[3]

Carbon atoms don't just make coal and diamonds, either. Arranged one way they become the graphite in your pencil, and in another "carbon nanotubes"—nanometer-sized tunnels which have become something of a poster boy for nanotech due to their astonishing properties. At certain diameters, carbon nanotubes conduct electricity with incredible efficiency. Arrange their atoms slightly differently and you get superefficient "semiconductors" (the key component in microchips) that generate far less heat and operate at much faster speeds than silicon-based counterparts. Another staggering property of carbon nanotubes is their strength—fifty times greater than steel—strong enough that a single strand the thickness of a hair could lift the weight of a family car.[4]

By controlling the atomic structure of just *one* element, we create a wealth of opportunities: superlight fuel-efficient aircraft, efficient transmission of power over huge distances, featherweight body armor, microchips running at speeds that make today's best look like Norm from *Cheers* competing in the Olympic 100-meter sprint, even a cable that could stretch all the way into orbit, providing a new route to space. And that's only carbon in its cape and bright red underpants. What might we achieve with other elements, both individually and in combination?

"The resulting abilities will be so powerful that, in a competitive world, failure to develop [them] would be equivalent to unilateral disarmament,"[5] Drexler has written. Guiding us through this change is, he says, "the great task of our time." Later he will tell me "the reason I set out in the direction

I did, and studied what I studied, is because I was convinced that civilization is coming to a carrying capacity crunch that could create a lot of suffering. I could see a way to change that. So I set out to save the world."

I'm nervous about meeting Eric for two reasons. The first is that he has impressed upon me that he's adopted a policy of not meeting writers face-to-face, preferring to only answer a few questions by e-mail. At his insistence I've nervously sweated over and sent him my questions in advance (on the basis of which he agreed to meet me), and I feel like I'm reporting to class. My second reason is that the "this technology will save the world" story is as old as technology itself and makes me naturally cautious. I'm worried I'm about to do the equivalent of looking at someone's baby and blurting out, "It's not as pretty as you think it is!"

But you can't understand the future without understanding nanotech— and Eric Drexler is to nanotechnology what Van Morrison is to folk music and drinking. Even his critics (of whom there are many) recognize his enormous contribution. Part of his reticence to meet is, I'm sure, down to the inordinately rocky ride that Drexler has been on since he burst onto the scene with *Engines of Creation*. This includes a high-profile intellectual boxing match with a Nobel Prize–winning chemist, and several other run-ins with senior scientists who've questioned his credentials, suggesting "he's an engineer who studied chemistry and thought about its possibilities, but perhaps did not take enough chemistry."[6]

Eric, you see, has a Big Idea—an idea whose first inklings came to him even before he wrote *Engines of Creation*, but which he didn't fully articulate until his second book, *Nanosystems*, in 1992. It's one that still divides science today, with researchers often referred to as "Drexlerian" or "anti-Drexlerian," which sound like the sort of words my long-haired dining companion might use.* In the course of promoting his idea, Eric's

---

*They are also terms that Eric rather hates. Attaching names to positions, he says, tends to give debates a personal edge, rather than remaining purely scientific.

been accused of unprofessionalism, shameless self-promotion, the fundamental failure to understand his own subject, that he writes under the influence of drugs and scares children (no, those last two aren't a joke). The subtext: "Drexler is crazy."

On top of all this, he's had many of his ideas quoted out of context, twisted by the lens of headline-hunting journalism and inaccurately appropriated for dramatic effect by science fiction writers. It's fair to say that over the years Drexler's Big Idea (and the man himself) have gotten under the skin of a lot of people. In 2004, *Wired* ran an article by Ed Regis (a largely pro-Drexler voice) with the strapline "K. Eric Drexler was the godfather of nanotechnology. But the MIT prodigy who dreamed up molecular machines was shoved aside by big science—and now he's an industry outcast."[7] No wonder he doesn't like talking to writers. I'm getting seriously worried that Eric Drexler is going to be *very prickly*.

The accepted history of nanotechnology says the field was born in a 1959 lecture entitled "There's Plenty of Room at the Bottom" by the late great physicist Richard Feynman. For a long time the talk remained a little-quoted curio, but it has gained prominence retrospectively largely due to Drexler popularizing it. Some people even suggested that Eric Drexler was trying to shore up his own credentials by associating himself with a "big name" in physics, which strikes me as a bit unfair.

During his talk, given to the annual meeting of the American Physical Society at the California Institute of Technology (Caltech), Feynman considered the possibility of writing the contents of the *Encyclopaedia Britannica* onto the head of a pin and concluded, if one could arrange atoms precisely to form letters, that not only could you get the *Encyclopaedia Britannica* on the head of a pin, you could get all the information that mankind had ever recorded in books in a small pamphlet. Taking the idea further, Feynman pondered how much material would be needed to store all the words in all mankind's books if those words were converted to digital code, and deduced that, with a good deal of room to spare (atomically speaking), all of the information accumulated in all the books in the world could be held in a cube of material one two-hundredth of an inch wide, "which is the barest piece of dust that can be made out by the human eye."[8]

It still sounds incredible today. Imagine how it sounded in 1959. And indeed it was thirty years later before we come to the next popular landmark in the nanotechnology story, when Donald Eigler became the first person in history to move and control single atoms in a repeatable and predictable way, arranging thirty-five xenon atoms on a bed of nickel to spell out the name of his employer: IBM.* "This capacity has allowed us to fabricate rudimentary structures of our own design, atom by atom," wrote Eigler and colleague Erhard Schweizer. "The possibilities for perhaps the ultimate in device miniaturization are evident."[9] In 1959, the most obvious example of storing information on the nanoscale was DNA (the discovery of its structure was the big news in science six years earlier). "That enormous amounts of information can be carried in an exceedingly small space is, of course, well known to the biologists," Feynman reminded his audience. Your DNA comes in strands just two nanometers wide, curled up inside the cell nucleus—a structure that is about seventy-five times smaller across than Feynman's dustopedia. Reading Feynman's lecture reminds me that my own body is a walking collection of billions of nano-sized manufacturing plants churning out cell stuff. Seen from this angle Harvard's George Church is a nanotechnologist too. If our cells can marshal materials on the nanoscale, then nanomanufacture is clearly possible. If it wasn't, you wouldn't be here.

In his lecture Feynman took his ideas to their logical conclusion and told his audience that there was nothing in the principles of physics that forbade the maneuvering of individual atoms. "It is something, in principle, that can be done; but in practice, it has not been done because we are too big."

Enter Eric, who is "too big" but very smart.

---

*In 2009, Eigler commented, "It was more than a publicity stunt. Emotionally, for me, it was much more important. This is going to sound hokey, but it's the truth. IBM picked me up off the scrap heap of science and gave me every opportunity a scientist could hope for to be successful. As far as I was concerned, it was payback time."

———

Eric Drexler arrives two minutes ahead of time, which gives me an excuse to move away from the guy with the long gray hair, who is now loudly holding forth about protecting against alien abduction by "visualizing yourself surrounded by gold energy."

We find a seat by the window and order cinnamon tea. Taking his first sip, Drexler remarks, "The tea is unusually strong right now," before adding, "it fluctuates." This is Eric Drexler in a nutshell, exact in every detail, expressing himself by marshaling facts as he sees them first, and only rarely (and carefully) expressing opinion. He doesn't say the tea is "good" or "bad" but remarks on its strength. "When I want to say that something is a good idea or a bad idea, desirable or not, that will be clearly flagged and said with qualification and trepidation," he tells me.

He's a wiry fellow ("Down by about forty pounds from the nineties!" he tells me), almost birdlike, with a well-kept gray beard that covers a long, thoughtful face, in which sit two piercing brown eyes that seem to say, "Yes? And what more?" Eric Drexler is not one to be satisfied with half a story, or abstract musings. He tells me he gets annoyed "when scientists suggest their work implies new insights but don't *explicitly state* what those insights are." You'd better know the history of your profession too. So, if you work for NASA and invite Eric for lunch but don't recognize the name "Konstantin Tsiolkovsky," he'll be mildly disappointed in you (as he tells me he was just a few weeks earlier). I hadn't heard of Tsiolkovsky either (but I don't work for NASA so I'm let off), but the reclusive Russian schoolteacher who was writing about multistage rockets for reaching Earth orbit as early as 1903 becomes a recurring theme in our conversation. Drexler sees little excuse for being ill-informed, educating himself on a wide range of topics (his blog contains posts entitled "How to Understand Everything—and Why" and "How to Learn About Everything"). In this sense he is mildly terrifying. But in person he's neither prickly nor intimidating. He's certainly exact, but he's generous too—and interested. He wants to know all about my trip to see "the remarkable

George Church" and is interested too in my investigations into robotics at MIT and Cornell. (His opinion is that "human-level Artificial Intelligence is inevitable.") Maybe his chatty and curious mood is because he's had a bit of a renaissance—Eric's Big Idea is back on the table—but I find nothing of the child-frightener about him.

Drexler's Big Idea is the "nanofactory"—a desktop-sized programmable device that, in principle, could manufacture, with atomic-level precision and absolute fidelity, any number of hugely powerful and useful things using a cheap supply of simple "chemical feedstocks." In the introduction to the twentieth-anniversary edition of *Engines of Creation*, he writes that

> a 10-kilogram factory will be able to produce 10 kilograms of products in hours or less—a stack of billion-processor laptops, a package containing a trillion cell-sized medical devices, or a roll containing hundreds of square meters of tough, flexible stuff that converts sunlight to electric power. It seems that raw materials will be the main cost of production. At a dollar per kilogram (a typical price for industrial feedstocks today) the solar-electric material would cost about one cent per square meter, and the computers would cost about a dime. Shipping and handling extra.[10]

By approaching matter in the way computers treat information, we move from a world where factories specialize in making one product to *programmable* factories, or "matter printers" that can make myriad products as long as they're fed the right design and raw materials—a multipurpose machine creating products to order at a fraction of the cost of today's methods. Just as computers "form complex patterns of the elementary building blocks of information," writes Eric, molecular manufacturing systems will "form complex patterns of the elementary building blocks of matter." Manufacturing becomes an *information technology*, a world where we can e-mail product specifications to each other, and our desktop "matter printers" make them.

"So, you're taking the Industrial Revolution model of the factory, putting it on my desktop, and I can program it to make, well, almost anything? It's Industrial Revolution Two-Point-Oh but computable?"

Eric thinks for a moment. "Yes. I think the comparison to the Industrial Revolution is a highly appropriate framing, not only because of the factory model, but also because when you change how things are made, and therefore what *can* be made, this has profound effects on everything."

One of those "profound effects" could be the end of capitalism.

I'd thought I'd left the *Star Trek* comparisons behind when I moved tables but if Eric's right then many of the ideas Gene Roddenberry explored in his iconic sci-fi series could come to pass. The argument goes something like this: How much we pay for things is based on two key factors, what they cost to make and how scarce they are. Nanotechnology has the potential to deal a sucker punch to both.

With Eric's nanofactory, raw materials are cheap chemical "feedstocks." "Common elements like hydrogen, carbon, nitrogen, oxygen, aluminum and silicon seem best for constructing the bulk of most structures, vehicles, computers, clothes and so forth: they are light and form strong bonds. Because dirt and air contain these elements in abundance, raw materials can be dirt cheap," writes Eric in *Engines of Creation*.

In this future, products are assembled by nanofactories following programs, not by human physical labor, so there's no need for wages. Need more capacity? Because the nanofactory is programmable it can quickly build a copy of itself (at a tiny cost too). Among the other products it can make, of course, are devices for generating energy, like superefficient solar cells—so the cost of energy becomes negligible. It's a world where nearly everything is cheap to make out of abundant raw materials. And if we really do need some esoteric scarce element we can't manufacture an equivalent for? Eric's answer is cheap nanomanufactured spacecraft that mine the solar system. The asteroid belt has resources of rare metals that make Earth's complement look like half a potato brought to the harvest festival.

It's such a fantastical proposition that my mind is scrabbling for a reality check. It's saying, "This is *too much*. I'm *being seduced*." Eric is talking about stuff that sounds as off-the-scale as all the talk of aliens and "prime directives" a few tables away. If William Shatner walked in now and announced *Star Trek* was actually a documentary, it'd somehow seem normal (and one long-haired bloke would probably explode with joy).

"It's pretty fantastical," I say. "Is it possible?"

Eric gives one of his exact and heavily qualified answers (my translation in brackets).

"I'm looking at phenomena we know exist in the world . . . [he means the laws of physics, chemistry and biology] . . . we don't yet know how all these work but that doesn't necessarily matter [engineers don't have to know everything about the chemistry of materials to build bridges]." He continues, "If you apply large design margins . . . [if you give yourself a nice big range of error] . . . you get a very reliable picture of what you can achieve if you actually had the resources—technology, money, people to put it together. It's a bit like Tsiolkovsky's rockets though—this might be rather far down the line [you need a *very big* research budget and a lot of patience]." In short, Eric is saying, "Yes, it is possible" and I'm getting dizzy.

Of all the things I've investigated so far, nanotechnology seems the closest to the realm of science fiction. In fact there's even a subgenre of sci-fi called "nanopunk" which riffs on future worlds changed by nano-engineering. José López from the University of Ottawa makes the point that science fiction and nanotechnology are deeply entwined, due to the field's "radical future orientation," which he argues "opens up a gap between what is . . . possible today and its inflated promises for the future."[11] It's a view I'm surprised to hear Eric partially agrees with.

"If anyone were trying to build a nanofactory today, I would say I have no idea what they think they are doing. It would make no sense. It'd be like trying to pick up rocks on the moon when we haven't yet put a satellite in orbit." He won't say when it's going to happen—just that it probably will.

On the way there, lots of things will change. Already, "nanoproducts" are coming to market which, if they live up to the hype, could radically reshape whole industries. Even if these first-generation products turn out to be marketing over delivery, the underlying science is sound, suggesting it's only a matter of time before they hit pay dirt. The first major casualty of the Nano Revolution?

Cleaning products.

We're used to thinking of cells as gloopy, semiliquid things that alter their shape as they bustle around each other, but not all cells are like that. Diatoms, a single-celled form of algae, have *hard* cell walls made of silica.* When diatoms die, they leave behind tiny glass skeletons staggering in their intricacy and beauty. These skeletons come in two parts, one half fitting snugly over the top of the other, like the two parts of a petri dish that children use to collect bugs. In diatoms, the sides of these overlapping parts contain tiny nanometer-scale holes, and by moving these two parts in relation to each other, a diatom can control how big an aperture it has on the outside world, allowing things it wants in (for example, molecules of oxygen) but keeping out things it doesn't (like bacteria). It's the tough, glass nanomembrane that keeps the diatom safe.

In recent years there have been numerous reports of manufactured silica coatings that mimic the diatom's defense mechanism. Like a diatom's cell wall these coatings contain tiny pores, nanometers across, that keep bacteria, dirt and viruses at bay, while at the same time letting the material underneath breathe. These coatings themselves are just one hundred nanometers thick, making them invisible yet flexible enough to cover anything from tanks to tank tops. Potential applications for this "liquid glass" include everything from battling infection in hospitals, to

---

*Everyday forms of silica include sand and quartz.

keeping monuments pristine, to protecting against rust, to making sure your clothes never get dirty. To clean anything you'll just need water. Goodbye, most detergents and cleaning products (and a whole industry with them). If it puts an end to annoying adverts for washing powder, then I'm all for it.

This is just one example of hundreds. Suddenly I'm seeing the "hand of nano" everywhere. A team at the University of California, for instance, is working on a technique that dispatches antibodies to sniff out cancer cells and then marks them with a gold "nanosphere." When exposed to a particular frequency of infrared light, the nanospheres heat up and literally bake the cancer cells to death. Following an injection of the antibodies "you could send a person home, have them shine a laser on the specific part of the body with cancer for a couple of weeks, and they could be cured," says professor Jin Zhang.[12]

Eric Hoek and colleagues at the California NanoSystems Institute are just one team working on new nanomembranes that could reduce the cost of desalinating water, solving the problem of predicted fresh water shortages.[13] And researchers at Stellenbosch University's Water Institute in South Africa have put carbon and nanofibres into common tea bags to create a bottle-based, impossibly cheap water filtration system that removes pathogens from dirty water sources—a potential boon for millions of Africans.[14] Nanomaterials are also about to see widespread use in the construction industry to "enable novel applications ranging from structural strength enhancement and energy conservation to antimicrobial properties and self-cleaning surface," according to a report from Rice University and the University of California.[15] In Greece, the Intelligent, Safe and Smart Built (ISSB) project is developing a house for earthquake zones that uses nanomaterials to self-heal cracks; nano-polymer particles that become liquid when squeezed under pressure, flow into the cracks and then harden into a solid material.[16] I could go on. The Project on Emerging Nanotechnologies keeps a watching brief on consumer products incorporating nanotechnologies. The list includes everything from computer memory to toothpaste, via medical dressings, guitar strings, every kind of vehicle, cosmetics and (it seems to me) a disproportionate number of hair-straightening devices.[17]

Another promising nanotechnology is "DNA origami," using DNA as *raw material* for making nanostructures (i.e., being used as a *structural substance* and not as a store of information). DNA makes a good building material because we know how it snaps together (A's love to bind to T's, and the G's only have eyes for C's). Paul Rothemund of Caltech used this two-way binding as a way to rivet bits of DNA together in interesting ways. One of his first demonstrations of this technique was to create a map of the Americas one 200-trillionth of their actual size.[18]

But there are more useful applications than "DNA nano-art." One example is building tiny nanoboxes with lids and "molecular locks" that might be used to contain drugs until they can be delivered to just the right cell—and then unleashed with devastating accuracy.[19] In another example, Rothemund and colleagues used DNA "staples" to carry some of those wondrous carbon nanotubes into position to form a tiny circuit switch even smaller than those found in modern microprocessors.[20]

In May 2010, chemists at New York University and China's Nanjing University used this DNA origami technique to create a nano-sized factory shop floor with an assembly line that could be followed by a "DNA walker" that picks up nanocomponents as it moves along the line. By the time the walker gets to the end, it has been added to, just like a car chassis slowly turns into a car as it makes its journey from one side of an automobile factory to the other.

There's already a nanotechnology revolution going on, but none of it looks like Drexler's desktop revolution (yet). Typing "nanotechnology and medicine" into Google returns me over three and half million hits; "nanotechnology and electronics" delivers a similar number. Professor Richard Jones, a onetime senior nanotechnology adviser to the UK government and author of *Soft Machines: Nanotechnology and Life,* writes,

I think the main impact of "nanotech" whichever flavor you think possible will be the reduction in cost of most manufactured products. This is already happening with electronic goodies (TVs, PCs, music systems, etc.). This same progression is likely to happen in biomedical and normal manufactured products as well (cars, houses, etc.). I don't think this means that everything will be free, just low cost.[21]

So, even cautious commentators like Jones (who rejects Drexler's nanofactory vision) predict a revolution.*

Unless, that is, we all get eaten by Gray Goo.

You can't mention nanotechnology without someone bringing up this particular version of Doomsday soon after.

In *Engines of Creation*, Drexler imagines tiny atomically engineered nano-devices called "replicators" that carry a program which instructs them to continually make more of themselves. He writes:

> the first replicator assembles a copy in one thousand seconds, the two replicators then build two more in the next thousand seconds, the four build another four, and the eight build another eight. At the end of ten hours, there are not thirty-six new replicators, but over 68 billion. In less than a day, they would weigh a ton; in less than two days, they would out-weigh the Earth; in another four hours, they would exceed the mass of the Sun and all the planets combined.[22]

---

*Less publicized but beavering away on nanoresearch just the same is the food industry—looking into new molecular tools for detecting and treating crop diseases and increasing the ability of plants to absorb nutrients. "In the near future nano-structured catalysts will be available which will increase the efficiency of pesticides and herbicides, allowing lower doses to be used," says the 2006 *Nanotechnology in Agriculture and Food* report by the European Commission–funded *Nanoforum*. In the same report, the idea of interactive food is discussed, a product that will "allow consumers to modify food depending on their own nutritional needs or tastes. The concept is that thousands of nanocapsules containing flavor or color enhancers, or added nutritional elements (such as vitamins), would remain dormant in the food and only be released when triggered by the consumer." Another idea is to create healthy versions of traditionally unhealthy treats—donuts, pretzels and cheesecakes that taste as bad as you like but with low levels of fat, salt or sugar.

The popular analysis was that we will be overtaken, eaten up by trillions of tiny mechanical robots, neither necessarily gray or gooey, but the name stuck.

It's easy to see how this worry became so fixed in the collective imagination. There's something "familiar" about this threat, and I think it's the similarity to the idea of pandemic, except this time it's an unstoppable "matter virus" rather than a new strain of flu. It's the ultimate "we've lost control of the machine"/"the machine is alive" horror story. Like the emotional trick I found my mind playing on me when I imagined George Church in a bug-destroyed wasteland, or imagining armies of intelligent robots taking over the planet, I can instantly picture the Gray Goo apocalypse—a faceless, inexorable blanket of uncaring, unfeeling destruction. There goes Parliament and the White House. There goes Swindon Town Football Club, all victims to an ever expanding techno-vampire sucking the life out of everything in its path, indifferent to the diversity before it.

Engines of Creation is replete with somber warnings like:

Dangerous replicators could easily be too tough, small, and rapidly spreading to stop—at least if we made no preparation. We have trouble enough controlling viruses and fruit flies.

And

How well we design ahead—and what we design—may determine whether we survive and thrive, or whether we obliterate ourselves.

In the book, Eric suggests that a wait-and-see policy could "cost many millions of lives, and perhaps end life on Earth." A lack of oversight, he warned, could render our future, like Kylie Minogue, "exciting and short." But Drexler's motivation for raising the Gray Goo issue was, he tells me, simple "civic duty"—a warning for us to be vigilant in case someone attempted to deliberately manufacture something Gray Goo–like. Meeting Eric gives me the strong impression there is nothing sensationalist about him. His catalogue of the potential benefits and downsides of nanotechnology is simply that—a catalogue with very little emotional weight attached to any entry. These are statements as controversial or emotional to him as "please mind the gap" or "the tea is unusually strong." Other people (understandably) didn't take it that way. In a bruising exchange of letters with Drexler, published in that well-known comedy periodical *Chemical and Engineering News*, Nobel prize–winning chemist Richard Smalley, famous for his work on those carbon nanotubes, wrote:

> You and people around you have scared our children. I don't expect you to stop, but I hope others in the chemical community will join with me in turning on the light, and showing our children that, while our future in the real world will be challenging and there are real risks, there will be no such monster as the self-replicating mechanical nanobot of your dreams.[23]

This kind of attack is something we're used to seeing in the arena of fear-based politics, but it's shocking to read it in an exchange between high-ranking scientists, especially when one of them has the Nobel Prize for Chemistry on the mantelpiece. It's also a shame that Smalley's memory (he died in October 2005) is rather tarnished (in some people's eyes) by his run-ins with Eric. He was one of his generation's finest chemists who is perhaps better remembered for something he and Eric agreed strongly on—the need for better science education and for which his strapline was "Be a scientist, save the world."[24]

The crucial fact that tends to get overlooked is that all self-replication (including that performed by cells) requires some method of collecting raw materials. If you stopped eating and breathing, your cells would soon cease dividing. Stop providing your body with fuel and it dies. Luckily (for your cells) they're part of a bigger system (you) that has the ability to nip to Safeway. (Try sending just your liver down to the shops and see how far you, it and all your cells get in the game of life.) Beyond that, for any nanomachine to self-replicate, it would need to have the ability not only to continuously *find* raw materials, but also free them from whatever they're already part of. "The link some writers made between my ideas of molecular nanotechnology and Gray Goo, are, frankly, a lot of crap," Eric tells me.

The second reason?

"Gray Goo makes a good scare story, as long as you leave out the science," says Eric. Indeed Michael Crichton's novel *Prey*, published in 2002, *is* this scare story, radically massaging the science in the service of drama.

"If you want a system that *manufactures* things, look at an automobile plant," says Eric. "Can it make another automobile plant? Not remotely. And an automobile can't make an automobile. Is there, in fact, any self-replicating system in a box? Has anyone set out to do it? No. Because there's no motivation. It doesn't work as a model of manufacture because it makes the whole thing staggeringly inefficient." Then Eric puts one of the strangest images of my trip into my head. "It's hard to design a car that gets pregnant and replicates itself," he says, and suddenly I'm imagining outraged owners returning to car parks to find a full-on automobile orgy in progress. (You'd *walk* home after that, wouldn't you?) Eric continues, "That's a difficult engineering challenge. It would be extraordinarily difficult to make, there's no reason to and the result would be very inefficient." In short, the world isn't built to support unstoppable self-replicating machines (on any scale) any more than it's built to support an evil strand of amoebas that could spiral out of control and eat up the rest of life on Earth.

Looking back at Gray Goo now, he says the idea is "stupid" and I get the impression he rather wishes he'd never mentioned it. Ironically, Gray Goo's most dangerous act of self-replication has been as an idea—and certainly one that contributed to Eric's spell in the cold.

From the outside and for a long period (almost twenty years) the field of nanotechnology looked rather like the brilliant kid who couldn't work out what to do with itself. "There was a bit of winter," says Eric, with a hint of sadness and an even tinier smidgen of resentment in his voice.

First, there was an identity crisis. "People entering the field found that, on the one hand, they were asked, 'So when are you going to deliver those tiny little submarines that'll clean up our blood?'" Eric recalls, referring to a recurring iconic nanotech image that is often trotted out by rushed newspaper picture editors and that most senior researchers regard as simply ridiculous.* "On the other hand, they were told, 'You're working on very dangerous things that are going to destroy us all,' which of course they weren't doing, either."

Second, former fans of Drexler were turning against him, creating the Drexlerian/anti-Drexlerian split that prevails today. Richard Smalley (who accused Drexler of scaring children in 2003) admitted that reading *Engines of Creation* "was the trigger event that started my own journey in nanotechnology,"[25] and even told the Congressional Hearings on Nanotechnology in 1999 that "there is a growing sense in the scientific and technical community that we are about to enter a golden new era. We are about to be able to build things that work on the smallest possible length scales, atom by atom with the ultimate level of finesse."[26] Looking back in 2004, he said, "After a while I thought I saw what might be some problems. The more I thought about it, the more troublesome they appeared. Finally I ended up thinking, it's just hopeless."[27]

At the root of the anti-Drexlerian position is what chemists *know* about how things fit together at the molecular level in nature, and what Drexlerian nanofactory advocates say is *possible* about how we might be able to fit things together if we accept only the laws of physics as our ultimate constraints. The chemists point out that when you get down to the nanoscale there's a mêlée of natural chemical reactions with molecules and atoms attracting and repelling one another and out-of-control physical collisions (for nanocomponents a rogue piece of dust is like an iceberg thousands of times bigger than the one that sank the *Titanic*). For them the nanofactory soon starts to look like me trying to make an airfix model of the *Starship Enterprise* when all its parts (and all my fingers) have

---

*To be fair to newspaper editors it's not only them using such images. Tiny little submarines have long been a staple of some pro-Drexlerian presentations too.

just been dipped in superglue and people are throwing bricks at me. While biology has certainly developed nanomanufacture, most notably in the form of cells, it's had to do it within the bounds of what chemistry will allow. Try to go too far against that, and chemistry will be knocking on your door like a cop with a search warrant. "Stop doing that, sonny—it's dangerous and it won't work, anyway." Every atom interacts with every other, including those that make up Drexler's nanofactory. For the chemist critics, the nanoscale looks far too much like the mosh pit at the front of a death metal gig for it ever to be a place to build an assembly line.

Eric counters: "People say, 'You're not thinking about the fact these molecular things stick together.' Well, that would be like saying someone who's designing aircraft isn't thinking about gravity. It's a fundamental part of the problem. If it *were* being ignored, that would be the same as saying that everybody in this field is utterly and thoroughly incompetent and doesn't know the first thing about molecules." He takes a sip of tea. "This turns out not to be true." The "it's all too messy" argument is clearly tiresome for Drexler. "The question isn't, Are there some things that won't work? The question is, Are there enough things that will?"

The early twenty-first century saw nanotechnology behaving a bit like a once-great rock band that had split up. Nanotechnology had done the same as the Eagles. But things were set to improve. Time was ripe for a reunion tour. After a long time in the cold, Drexler's nanofactory idea was about to have a partial renaissance.

In 2006 a joint report from the National Academy of Sciences, the National Academy of Engineering and the National Research Council concluded that "manufacturing processes with some capability to pattern structures with atomic precision" might one day be possible, and research into programmable manufacture at the molecular level should be cautiously encouraged. That report's final paragraphs are carefully worded, pointing out that the future performance of many theoretical components of molecular manufacturing cannot be "reliably predicted at this time." Nonetheless, they're clear that the theoretical feasibility of a Drexler-like system cannot be argued against. To make the point, they reference the now familiar Russian schoolteacher and rocket fan. "This work is currently outside the mainstream of both conventional science (designed to seek new knowledge) and conventional engineering (usually concerned with the design of things that can be built more or less immediately). Rather, it may be in the tradition of visionary engineering analysis exemplified by Konstantin Tsiolkovski's 1903 publication, 'The Exploration of Cosmic Space by Means of Reaction Devices.'" In other words, "Watch this space."

Let's assume for the moment that Eric's ambition is possible—that at some point in the future the nanofactory is coming. The implications are enormous. Humans will no longer have to make stuff. Manufacture will be a design-only profession. What you can imagine (within the laws of physics) will be makeable, including machines that generate almost free, clean energy, and computers with calculational powers that will appear almost magical, millions of times faster and millions of times more energy efficient than their counterparts today. Manufacturing capacity will be, essentially, unlimited. New products will be able to be prototyped within days rather than years. Ideas can become matter. We will want for nothing.

Then again, there is the end of the world to think about. There are at least five versions of a nanotech-made apocalypse.

Nanotech apocalypse 1: Economic meltdown. One nation gets the nanofactory ahead of everyone else. Suddenly they don't need to buy very much anymore. So, for instance, the United States stops buying goods from abroad. Economies dependent on U.S. trade go into a tailspin, resulting in untold suffering and igniting a global economic crisis.

Nanotech apocalypse 2: Nanotechnology arms race. Nations terrified of the threat posed by foreign powers who might possess this technology ahead of them race to produce their own version, driven by the same desires that saw nuclear arms proliferation. Once again, Mutually Assured Destruction via powerful nano-engineered weapons is on the table and the world lives in fear of or, worse, suffers an all-out nano-war.

Nanotech apocalypse 3: Welcome back, Hitler. At the giggle fest that was the 2008 Global Catastrophic Risks conference, cofounder of the Center for Responsible Nanotechnology Mike Treder said, "If you can make any product you want on your desk in your home, what happens to all the people whose jobs rely on finding raw materials, manufacturing those products, delivering them—transportation, storage, wholesale, retail—all of those people might someday be left jobless. . . . Such social disruption can leave a vacuum that allows a powerful charismatic personality to take hold. We've seen that happen before . . . and it could lead to a situation of global tyranny." In short, traditional power structures fail under the weight of technology's advance and the next version of Adolf Hitler uses the confusion to take over.

Nanotech apocalypse 4: The democratization of violence, Says Mike, "The ability of smaller groups or even individuals to wield greater destructive power. If any technology was ever primed to do that this is one." In short, ideological nutters have the ability to kill a lot of people because their nanofactory can manufacture weapons of unimaginable destruction.

Last but not least is Nanotech apocalypse 5: Police state. Nanotechnology allows you to be watched without your knowing. A billion unseen cameras, microphones and tracking devices could be used to observe and then suppress a populace, making McCarthyism look like your neighbors twitching the curtains. "More effective, more ubiquitous, more invasive surveillance," says Mike. This is Eric's personal "favorite." "That's what people should be scared of," he tells me. "Not terrorism but the ability to *suppress anything like terrorism.*"

And of course, these could all combine in some unimaginable way. Nanofactory-inspired global economic meltdown and social disruption lead to the rise of dictators who inherit the fruits of a nanotech arms race and fine-grained surveillance. We settle down into a world of fear under oppressive regimes where you can't go to the toilet without it being reported (if we haven't all been killed in the process).

Oh, dear.

The problem with all these scenarios is they're naturally built on our understanding of history. We're using a lens shaped by the past to look at the future. But with a nanofactory revolution, the lens is just wrong. It can't help us see a world without scarcity because we've never had one. If we learn to control matter in the way Eric suggests, then we're making a shift as fundamental as our primate ancestors did when the idea of using tools first occurred to them. As primitive as using a jawbone for a club looks to us now, so industrial manufacture will look to our descendents (or maybe us, if it comes soon enough). If you went back to a caveman and tried to explain the agricultural or industrial revolutions to them they might respond with "so how many bear skins and jaw bones does this mean I get?" Similarly when we're presented with the idea of the nano-factory revolution we might respond with "will this affect my pay rise?" Any nano apocalypse we imagine using our old lens is also equally flawed.

Two things seem clear. First, the Nano Revolution won't be owned by any one nation. Nanotechnology is an international discipline. The examples in these pages come from across Asia, the United States, Africa and Europe. There are nanotech initiatives and hubs in the Middle East, India and South America. There is no dominant party. Second, it won't happen overnight. As I write, it's nearly a quarter of a century since Drexler laid out his vision in *Engines of Creation* and Eric himself admits anybody trying to build a nanofactory would be entirely misguided. Just like Konstantin Tsiolkovsky and his rockets in 1903, we're still a long way off having the engineering to get us from here to there. Nanotechnology will enter our lives in increments (and influence other industries in increments too). The world will not turn on a penny.

Taking these two observations together I find it hard to imagine either a nanotech-inspired economic meltdown or a scenario where one nation suddenly acquires the nanofactory and then either a) suppresses all knowledge (worldwide) related to nanofactory development, or b) uses the technology to suppress/invade other territories quickly enough to ensure some kind of draconian order. The other reason a war brought on by nanotech is hard to imagine is answered when you think about the question "Why do we fight in the first place?"

The most convincing theory I've come across is related to the Earth's "carrying capacity" that Eric mentioned early in our conversation. In short, human beings have a habit of outstripping their natural environment, and when they do, they start to compete for what have become limited resources. We want more land, food, minerals, labor or fuel, and suddenly there isn't enough. Nations try to corral sufficient resources and sometimes that means taking them from someone else. As the population rises we have two choices: find a way to increase carrying capacity, or fight (justifying the latter in a way we feel comfortable with, usually some form of ideology). One of the great worries surrounding climate change is that it might radically reduce the carrying capacity of the Earth, leading to wars over resources exploding across the surface of our planet. Of course there are other reasons people go to war, but on balance theories concerning the causes of war, whether from economic, evolutionary or demographic perspectives, tend to stress that competition for resources is a dominant part of the mix.

But what has this got to do with nanotechnology? The same Stewart Brand who suggested the impacts of nanotechnology would be "revolutionary-times-revolutionary" provides a link in his book *Whole Earth Discipline: An Ecopragmatist Manifesto*:

> Peace can break out, though, when carrying capacity is pushed up suddenly, as with the invention of agriculture, or newly effective bureaucracy, or remote trade or technological breakthroughs.[28]

Nanotechnology can potentially (and dramatically) increase carrying capacity and, crucially, the distribution of resources—because everyone has the raw materials needed to make whatever they need, including food. The Food and Agriculture Organization of the United Nations states that "world agriculture produces seventeen percent more calories per person today than it did thirty years ago, despite a seventy percent population increase"[29] and yet in recent years the number of undernourished people has risen.[30] Nanotechnology has a powerful role to play in reversing this trend.

And, as Eric maintains, at the same speed at which nanotechnology offers us more inventive weapons, it's offering a reason not to use them. "You have an inconsistent scenario where you're developing a technology to fight wars that there is no reason to fight," says Eric. "So military planners need to understand that. To the extent you buy into this technology you're buying into a transformation. It's a different world. Scenarios where the motivation is to win resources are incoherent." This makes Nanotech apocalypse 3 (welcome back, Hitler) difficult to imagine too. Dictators require some inequalities upon which to sell their ideologies, but technology tends to reduce inequality. In fact, it's equally valid to argue that the greatest risk to society is not *getting* the nanofactory, but not getting it *soon enough*.

Reflecting on the other two concerns—ubiquitous surveillance (police state) or putting extreme weapon-making power into the hands of idiots—I'm transported back to Harvard with George Church, talking about the power of synthetic biology and the need for a scientist-led movement for industry oversight and licensing. Nanotechnology is a younger science and is only just waking up to the need to work out a regulatory and ethical framework that brings us the benefits, but avoids the pitfalls. While nanotechnology has the potential to reduce the biggest threats mankind faces, most notably by increasing the planet's carrying capacity, as yet it has no credible roadmap of oversight and ethics. This is worrying, but not uncommon. Regulation tends to get off the ground after we've had a few accidents and outrages; and nanotechnology, like synthetic biology, will have its fair share. The "good" news is that those first accidents will be as much a reflection of the technology's advancement as its successes.

There is no chance of your crazy flatmate getting hold of a nanofactory and vaporizing the neighborhood with a new nanoweapon anytime soon. As for ubiquitous nanosurveillance, there is a black hole in legislation that also needs to be addressed as the technology matures. That said, surveillance and antisurveillance technologies also develop in step. Today you can buy cheap antispy camera scanners that let you know if there's a device transmitting wireless camera signals in your vicinity and find out where it is (either by homing in on its signal or using LEDs to reveal the glint of hidden camera lenses).[31]

In 2004, the Royal Society and the Royal Academy of Engineering summed up the balance between promise and peril nicely in their appropriately named report *Nanoscience and Nanotechnologies: Opportunities and Uncertainties*:

> The convergence of nanotechnologies with information technology, linking complex networks of remote sensing devices with significant computational power, could be used to achieve greater personal safety, security and individualised healthcare and to allow businesses to track and monitor their products. It could equally be used for covert surveillance, or for the collection and distribution of information without adequate consent.[32]

I suspect therefore that nanotechnology will follow synthetic biology's example. As the science becomes more "real," regulation will evolve, and some people will die because we didn't work out all the risks in advance. There will be headlines. Already there are worries that carbon nanotubes are "the new asbestos." It is concerns like these that led the Royal Society to suggest that "nanoparticles and nanotubes should be treated as new chemicals under UK and European legislation, in order to trigger appropriate safety tests and clear labeling"[33]—a strategy that seems to characterize many legislative responses around the globe. But nobody is legislating for the nanofactory because it's still massively theoretical.

Does Eric have an idea about when we should *start* worrying about the legislation? How far off is the nanofactory revolution? "I'm not at all inclined to talk about time frames," he says. "What I will say, is that the scientific knowledge and the tools are in place such that with a lot of creativity we can get beyond *some* of the difficulties of designing and using molecules in the near future." In other words, "don't hold your breath but we're working on it."

But in some ways it doesn't matter if Eric Drexler is right or wrong. Professor Richard Jones makes the point rather brilliantly:

> Followers of Drexler are in danger of finding themselves in denial about the potential impact of ordinary, evolutionary nanotechnology, because of their devotion to their brand of nanotechnology's one true path. . . . It would be ironic if, in thirty years, the Drexlerites find themselves still waiting for a revolution that's already happened.[34]

Even if we don't get as far as the nanofactory, nanotechnology will continue to be exploited in energy and food production and in creating new materials. Jones suggests we

> shouldn't abandon all of the more radical goals of nanotechnology, because they may instead be achieved ultimately by routes quite different from (and longer than) those foreseen by the proponents of molecular nanotechnology. Perhaps we should thank Drexler for alerting us to the general possibilities of nanotechnology, while recognizing that the trajectories of new technologies rarely run smoothly along the paths foreseen by their pioneers.[35]

Or, "Maybe Drexler is crazy, but thanks."

My own feeling is that Drexler isn't crazy, but he is a dreamer. (That's not an insult. Martin Luther King Jr. didn't stand on the steps of the Lincoln Memorial and say, "I have a plan.") Whether Eric turns out to be right or not, he'll be remembered as the man who got the nanotechnology ball rolling, and that alone may well realize his ambition to "save the world." In the near future, nanotechnology looks set to underpin revolutions in energy production, medicine and sanitation. It seems, depressingly, that politics has rather obscured the revolution. But whether nanotechnology gets over its infighting or not, it has strong potential to make our world cheaper, healthier, more sustainable and a lot cleaner.

7

"The Biggest, Baddest Kid on the Block"

> After one look at this planet any visitor from outer space would say "I want to see the manager."
>
> —*William S. Burroughs*

'm on an eight-hour train ride that will take me deep into the California desert. As the scenery rolls past, I can feel my mind trying to stretch to accommodate the idea of societies remade by nanotechnology. Once more, I find my long-held views about what the world is and how it could be are being challenged. It's invigorating but tiring. I'm less than halfway through my journey, but already I feel like I need a holiday.

But the mind-bending is far from over. My next stop will challenge even the idea of what a holiday can be. . . .

I'm going off-planet. Which is what brings me to a desert town and the lounge of a roadside bar. Ozzy Osbourne is playing loudly over the juke-box, too loud for me to concentrate on my book. There are no windows.

Finishing my dinner, I find a seat at the bar in the adjoining saloon. Ozzy isn't quite so obtrusive here. There are two other customers, silent in contemplation. I order a beer and start reading.

"Never seen anybody read a book in here before," says the bigger of the two men.

I'm beginning to get the feeling that this isn't the kind of bar that holds weekly philosophical meetings.

"I don't have to read," I say, turning to him. "Let's talk."

Traveling alone gives you a hunger for conversation.

"You live here?" I ask.

"In the bar?"

"No, Mojave. The town."

"Sure," he says and falls silent, mulling over his beer.

"Were you born here?" I ask, persevering.

"No, I just moved from Los Angeles."

"That's a hell of a change."

Mojave, it has to be said, is hardly a thriving metropolis. It's essentially one road with a McDonald's, a Carl's Junior, a Burger King, a Denny's, a gas station and a clutch of motels along its length. The whole town is not much more than three blocks wide, blocks sparsely populated with shabby-looking homesteads. The social scene, with the exception of Mike's Roadhouse Café and Bar, seems to be based entirely around church (I've counted six) and a thrift shop. There's a coffee and donut hut opposite my motel, with a police car regularly parked outside.

"Why did you move?" I ask.

"Less trouble here," he says resolutely.

"You had trouble in Los Angeles?"

"I was in jail."

Now I understand. I'm in a road movie from the seventies. Any moment now Clint Eastwood is going to walk in and there will be a gunfight. I'll dive behind the bar as bullets shatter the bottles above me.

"Can I ask what you were in jail for?" I inquire, trying to sound nonplussed.

On the subject of violence he becomes a bit more vocal.

"Attempted murder. I bashed someone over the head during a robbery."

To my knowledge this is the first person I've ever met who's been to jail for trying to kill someone. Clint is late. Where is Clint?

"I needed money for drugs," he explains, warming to his theme. "I don't do drugs anymore. There's a lot less drugs in Mojave."

To be honest, there's a lot less *everything* in Mojave, with one notable exception.

Just over a hundred years before my journey to this town in the heart of the California desert, two brothers chose the similarly remote location of Kitty Hawk, North Carolina (nearly three thousand miles east of here) to test out their new machines. Relative isolation freed them from the prying eyes of reporters and potential competitors. They were neither wealthy nor government funded, but what they did changed the world. The brothers in question were bicycle engineers Wilbur and Orville Wright, who, late in 1903, achieved controlled, powered flight. Today, over three million people fly every single day.

I've come on a three-day trip to Mojave to meet some of their modern-day equivalents. Mojave may be short on fine dining and excuses for drug-inspired attempted murder, but it does have an excess of spaceships.

On April 12, 1961, Russian cosmonaut Yuri Alekseyevich Gagarin became the first man to orbit our planet, his achievement part of a geopolitical "superpower" space race that has dominated the tone of human spaceflight ever since. Human-carrying space missions are a matter of national pride and ego, open only to a select few. In the nearly fifty years since Gagarin's flight, less than 550 people have had the opportunity to follow his example.

In 1968, the film *2001: A Space Odyssey* imagined manned missions to Jupiter, hotels in orbit and a spaceflight industry that looked like an extension of the aviation business. It seemed like a fair prediction at the time. The same year, as Frank Borman, James Lovell and William Anders orbited the moon in *Apollo 8*, Pan American airlines started taking the names of passengers for commercial lunar flights. Why not? After all, it took just thirty-six years from the Wright brothers' breakthrough to Pan Am offering a luxurious transatlantic passenger service between New

York and Marseilles.*[1] Surely by the next millennium we would be taking similar trips to the lunar surface?

Paradoxically, the roots of our failure to open up space are the same roots that led to our early successes in traveling over the Kármán line (the point at which the air gets too thin for normal aircraft to fly, the point at which you need a *spacecraft* to keep going). The space race wasn't driven by a desire on the part of Russia or America to usher in new technological, commercial or humanitarian possibilities. It was a militaristic bun fight. "Apollo was never designed to open up space—it was designed to make us look like the biggest, baddest kid on the block," says Rick Tumlinson,[2] cofounder of the Space Frontier Foundation, which believes it is possible to achieve "large-scale industrialization and settlement of the inner solar system within one or two generations," an ambition they believe has been stifled by a "centrally planned and exclusive U.S. government space program."[3] For Tumlinson, our failure to grant widespread access to the solar system is due to politics, not possibilities.

Ever since the Apollo missions, NASA has dominated the debate and popular understanding of space, reinforcing the notion that human spaceflight is so costly and risky it is only achievable with huge levels of public funding. (By its own estimates the average cost of launching the Space Shuttle "is about $450 million per mission."[4])

Space exploration has become a poster boy for vainglorious excess in a troubled world. Patrick Stewart, who played *Starship Enterprise* captain Jean-Luc Picard in *Star Trek: The Next Generation*, summed it up when he told the BBC he worries the cost of human exploration of space "would take up so many resources, which I personally feel should be directed at our own planet."[5] It's not that there haven't been tangible benefits for humanity in our exploitation of the stars (including better understanding of our climate systems, reduction of fossil fuel consumption thanks to GPS routing, global communications networks and the development of

---

*By the time Pan Am went bankrupt in 1991, it reportedly had 93,000 names on its lunar reservations list.

new drugs and materials stemming from experiments in low gravity) it's just that for many people the price tag and the rewards have a less than direct correlation. Are we getting value for money?

But what if access to space became radically cheaper? What if space started to deliver real value-for-money in the eyes of the world? For instance, the moon is an abundant source of Helium-3, a nonradioactive form of helium that has been proposed as a clean nuclear fuel. The asteroid belt is awash with precious minerals. What if we could put an end to many resource shortages by mining the lifeless rocks of the solar system? In 2001, NASA's Near-Earth Asteroid Rendezvous (NEAR) mission landed on the asteroid Eros. Analysis of this twenty-one-mile-long, eight-mile-wide and eight-mile-thick hunk of rock revealed it contained "precious metals worth at least $20,000 billion."[6] (Of course if all that aluminum, gold and platinum ever made it to Earth, they would flood the markets, essentially ending any short-term scarcity, so the twenty-trillion-dollar valuation would instantly become redundant. The metals wouldn't be scarce anymore, reducing their price accordingly.)

Others suggest that if we could afford to get more of our world leaders to see the Earth from space, they might embrace environmentalism and see the value of increased cooperation among nations. Apollo 11 astronaut Mike Collins said, "The overriding sensation I got looking at the Earth was, my God, that little thing is so fragile out there,"[7] while Apollo 8's Frank Borman told *Newsweek* that looking back at the Earth brought home the thought that "this really is one world and why the hell can't we learn to live together like decent people?"[8]

For some, space is not just about resources or inspiration, it's about survival. At some point the Earth will be consumed when the sun goes supernova. Before that, there is the risk of a cataclysm befalling us, such as the asteroid that wiped out the dinosaurs. Science fiction author Larry Niven once said, "The dinosaurs became extinct because they didn't have a space program. And if we become extinct because we don't have a space program, it'll serve us right!"[9] (In June 2010 NASA announced that its Kepler space telescope had found candidates for seven hundred and six Earth-like planets after surveying one hundred and fifty-six thousand stars in a single patch of sky.)[10]

But whatever your view on the possible reasons for going to space—whether it's to find resources or inspire environmental awareness—the sticking point is cost. I've come to Mojave to meet the entrepreneurs and engineers who are hoping to solve that problem.

The sign says "Mojave Air and Space Port." If the term "space port" conjures up a futuristic image of gleaming towers, think again. This is a sprawling industrial park populated by nondescript sheds and factory units. As airports go, it's on the shabby end.

Still, it was here that Richard Branson's Virgin Galactic grabbed headlines in 2004 by launching *SpaceShipOne*, winning the $10 million Ansari X Prize for Spaceflight, offered to the first commercial vessel able to take three people over the Kármán line and return them to Earth twice in fourteen days. The total cost of the *entire development* was rumored at between twenty and thirty million dollars, which as far as the history of human spaceflight goes is peanuts. It's worth stressing that *SpaceShipOne* was a "suborbital" vehicle that took only short hops into space, but nonetheless the price tag was, by previous standards, ridiculously low. Compare it to the $330 million list price of an Airbus A380[11] (of which Virgin Atlantic have six on order[12]).

SpaceShipOne, now hanging in Washington's Smithsonian National Air and Space Museum, was built by Scaled Composites, whose operations are based here.* They're now working on building a small fleet of its successor, the unsurprisingly named *SpaceShipTwo*, which will take six passengers on two-hour flights beyond the lower atmosphere for a cool $200,000 a ticket. All told, the whole Virgin Galactic business is built on an investment less than the cost of one airliner or a single shuttle launch.

**SpaceShipOne hangs between the *Spirit of St. Louis* (the first plane to fly nonstop from New York to Paris) and the *Bell X-1* (the first controlled aircraft to exceed the speed of sound).

Scaled Composites isn't the only designer and builder of commercial spacecraft in Mojave. I walk past hangars for Masten Space Systems and XCOR Aerospace, too. XCOR's CEO Jeff Greason (whom I'll meet later) was part of the 2009 Augustine Commission that reviewed the workings of the American space program. Their report made the case for stronger partnerships between NASA and private spaceflight operators. Greason later remarked that their findings departed from previous presidential commissions on the future of U.S. space exploration not by recommending anything radically different but because "some people actually paid attention this time."[13] Part of the reason, no doubt, is the Space Shuttle's 2011 retirement date. With no launch vehicle immediately to replace it, NASA must look to commercial providers to get its hardware and astronauts into orbit (most notably to staff the International Space Station). Another reason is that by winning the X Prize, *SpaceShipOne* provided a real-world demonstration of cheap commercial spaceflight in action, grabbing the attention of the nation's accountants.

A number of commercial spaceflight companies have popped up in the last decade, most funded by billionaires looking for a new challenge and/or some ego food. Some are beginning to do serious business. Los Angeles–based SpaceX (started by PayPal founder Elon Musk) didn't even exist at the turn of the century but has already secured launch contracts from NASA and commercial satellite developers. The company claims it has developed all the flight hardware for its Falcon 9 orbital rocket, the Dragon spacecraft that is designed to sit on top of it, as well as three launch sites for less than the cost of one of NASA's launch towers—all while making a profit.[14] It hopes to be ferrying passengers to the International Space Station in the coming years.

Other start-ups are less interested in launch vehicles and more in spacefaring living spaces. Hotel entrepreneur Robert Bigelow already has two "inflatable" test habitats circling the planet and hopes to put a chain of space hotels and commercial residences into space, on the surface of the moon and, eventually, Mars. Before leaving for America I'd spent a mindblowing afternoon with Xavier Claramunt of Barcelona-based Galactic

Suite, who claims he'll have his first two paying guests in a space hotel by 2012.

"Our philosophy is the socialization of space," he'd told me over gin and tonics in the bar below his office. "In fifty years we'll have sixty hotels around the world in Low Earth orbit. Space is nearer than people know. In the next fifteen or twenty years, space will be like going to the seaside. I am sure *you* will go to space."

Xavier is also team leader of the Barcelona Moon team competing for the Google Lunar X prize, an international competition for privately funded companies "to safely land a robot on the surface of the Moon, travel five hundred meters over the lunar surface, and send images and data back to the Earth" (a competition in which over twenty teams are taking part).[15]

There's a new excitement about space again, and a lot of it has been generated by people working in this dusty collection of huts and hangars deep in the California desert. As Jeff Greason of XCOR says, "Mojave is to the emerging space industry what Silicon Valley was to the computer industry. There is a concentration of like-minded businesses here like nowhere else in the world."

My first stop is to see Stuart Witt, general manager of the space port. I'm hoping he can give me a measured view of the burgeoning commercial spaceflight industry. Like any frontier, commercial spaceflight is characterized by big personalities willing to take large risks. Business and egotistical motivations become easily entwined, and the hyperbole can become intoxicating.

Like many young boys, I was obsessed with space, determined that I was going to be an astronaut. When a visit to the optician's saw me leaving with spectacles, I understood my dream of spaceflight was over. I'd devoured every book I could on the subject of space exploration and knew that to get into space you had to be a pilot ace first, and pilot aces needed perfect eyesight. It took me at least a year to get over it. In some ways I'm not sure I ever did. Being here has reignited those boyhood passions. The promise that a commercial space industry may one day make it possible for me to see our planet from orbit is too delicious for words.

That's why I want to talk to Stuart. He's well respected in the industry, knows most of the players in it and provides services to many. He's one of the founders of the Commercial Spaceflight Federation (a body for discussion of standards, best practices and safety) and as such is not wedded to any company's particular story or ambitions. I feel I need someone who can help fortify me against any overly optimistic dreams of leaping on an economy flight to Earth orbit in the near future. In the alternately secretive and PR-heavy world of commercial spaceflight, I suspect he'll have a balanced view.

Stuart's office looks out onto one of three runways, and during our conversation light aircraft take off and land at regular intervals. On the walls are signed pictures of astronauts and aviators, interspersed with hunting paraphernalia. Stuart takes the chair opposite me, which is draped with an animal hide.

"You're a hunting man?" I ask.

"The only red meat my wife and I eat is what I kill, the only fish we eat is what I catch," he tells me. "It's not more expensive, it doesn't have any junk in it, it's tastier and this elk"—he pats the hide—"enjoyed a higher quality of life than anything that ended up in a supermarket."

An ex-navy pilot, Stuart has a no-nonsense manner about him. He approaches conversation the way combat pilots approach missions—find the target and eliminate it. If he wants to make a point, he won't pussyfoot around. This isn't to say he's rude, in fact he has a genial charm. Just don't waste his time. If you've got something to say, say it. Disagreement is O.K., but if he thinks you're bland you might as well turn right around and head home. He tells me, "The only people I care to be around are the ones that *truly live* the life they're given."

The efficiency of his words is matched by his facial expressions too. He can somehow signal approval or disdain just by altering the shape of his lip. Perhaps this is why his face is so smooth for a man in his mid-fifties. Wrinkles haven't had much of a chance to form. Overall the impression is of a stocky eight-year-old kid who just got bigger physically, but not that much older.

It turns out we had childhood dreams in common, but Stuart had better eyesight than me. He went through "Top Gun" and flew fighter jets for the U.S. Navy. His ultimate aim was to get to space, a dream punctured when he applied for astronaut training at thirty and was turned away. While Stuart had "the right stuff" in an airplane, he didn't have it in the classroom. "I didn't have enough science," he says.

He still intends to get to space, but he's cautious about the time frame.

"The risk the marketing people have now is falsely setting the expectations of the buying public. Space is still difficult. I would like to see hundreds of thousands of people get to space because I want to be one of them, but it's not going to happen overnight."

First off, the industry has to deal with the issue of safety. The space shuttle has a fatality rate of just under two percent. That sounds low, but if commercial airlines operated under similar conditions, over *fourteen million* people would have died in air crashes in the United States in 2009 alone. The commercial aviation industry, in cooperation with its regulators, has achieved the staggering feat of taking what looks like one of the most dangerous forms of transport and making it the safest. According to the U.S. Department of Transportation, flying is over five times safer than recreational boating and over thirty-five times safer than walking.[16]

Advocates of commercial spaceflight argue that the same will happen in their industry, but it won't be an instant transformation. The early years of commercial aviation saw levels of fatalities we would consider unacceptable today. The average number of accidents per million miles flown was 375 times higher between 1927 and 1937 than it was between 1996 and 2005.[17] Of the first nine de Havilland Comets (the world's first commercial jet airliner) built, five crashed: two failed to get off the runway in time, one broke up in a storm over India, one ditched in the Mediterranean and another plummeted into the sea shortly after takeoff from Rome. "Welcome on board. In the very likely event of a crash, at least your family will be able to say you died doing something glamorous."

Mojave has already witnessed tragedy. Less than a hundred yards from where I'm sitting, in the space port's Legacy Park is a memorial to three employees of Scaled Composites who lost their lives in July 2007 taking part in a propellant flow system test for *SpaceShipTwo*.

"At this facility I give you permission to kill yourself," says Stuart. "Here's my only limit on that: I don't want you taking out the whole neighborhood when you do it."

Scaled Composites' Burt Rutan, a world-renowned aerospace designer, is clear. "Don't believe anyone that tells you that the safety will be the same as a modern airliner."[18]

The question isn't *if* there will be an accident, it's when.

What that means is, if you want to take a trip into space in the next ten years, not only will you need somewhere between a hundred and two hundred thousand dollars spare for a ticket, you'll have to sign an "informed consent" document. This will be a more intense version of the piece of paper that dangerous-sports companies get you to sign before they'll let you get in a go-cart/go whitewater rafting/get a lift with Lindsay Lohan. You know, the form that says, "You could end up losing a limb or worse if you do this. Sign here to promise you won't sue us if you do." This is the only way such firms can get insurance, and the baby commercial spaceflight industry will be no different.

"The people who will regulate this industry will be the underwriters," says Stuart.

Risk is part of the deal here, and part of the attraction. Like everyone I meet at Mojave, Stuart has a deep-seated belief in the need for human beings to push boundaries and to explore. Commercial ambitions are subsumed into a higher aim than a simple search for profit. Pretty much everyone I speak to here sees access to space as a key battlefront for saving the human soul.

"We are creatures that require exploration, it's in our DNA," he says. "Humans require risk-taking, they require bold leadership that is willing to investigate the things they don't know, because every society that has gone safe has fallen."

In this place, asking the question "Why should we go to space?" would go down about as well as asking the Queen to do a *Playboy* pictorial.

"I was raised out west where the night sky was this gorgeous canopy of stars and you could dream big," says Stuart. "Some of us have a burning desire to be part of something greater than ourselves. That's the reason I'm here. I don't see that as a conservative or liberal viewpoint, it's who we are as a species."

I don't know how he manages it seemingly without moving a muscle, but Stuart's face is now alive with a kind of fire.

"Innovation lives here," he says urgently. "We have unrestricted dream-space. That's what we believe in."

Suddenly I'm eight years old again, dreaming of spaceships and touching the heavens, wanting some of that "unrestricted dream-space" for myself. This may just be an industrial park in a sandy field on one side of a sleepy desert town, but something is happening here that will change the way we see ourselves. *Homo sapiens* are finally going to get into space in larger numbers. We will become a space-faring species.

"Want an Elk Burger?" says Stuart.

Over lunch in the space port's café Stuart introduces me to Dick Rutan, one of the world's most respected test pilots. He's perhaps most famous for copiloting *Voyager*, a plane designed by his brother Burt to fly around the world nonstop without refueling. The journey took nine days and covered 26,358 miles on one tank of gas.[19] For aviation buffs, Dick is a superstar. He's aviation's version of Bono. Two people ask him for his autograph as we are talking.

Dick has a history of putting himself in dangerous situations. He's been shot down over Vietnam, crashed in the Arctic and had to eject over the English countryside (while worrying he was going to plow his failing aircraft into the town of Brandon, Suffolk).

"I had about a second and a half," he says. "If I ejected in that second and a half, I could live; if I ejected before or after, I would die; so I didn't have a whole lot of time. I gave the airplane a little bit of left trim to avoid the town as I was reaching for the ejection panels."

Rutan is like a coiled spring. Despite being over seventy he's still flying high-performance aircraft and you get the impression he'll be doing it in another seventy by sheer force of will alone. He swears prodigiously, almost spitting out the expletives.

"Stephen Hawking is a prick!" he announces. "You're telling me I can't go faster than the speed of light yet there's a gravitational force in a black hole that's strong enough to stop light and turn it around? Bullshit! Einstein's theory of relativity? Shame on you! Bullshit!"

Believing in limits is heresy here.

"Never look at a limitation as something you ever comply with. Never. Only look at it as an opportunity for greatness."

Listening to Dick Rutan is compelling in the same way watching a car crash is riveting. You kind of want to look away but something akin to morbid curiosity gets the better of you. I don't mean to imply Dick Rutan is some kind of psychological wreck (his mind is as agile as the planes he's asked to test). More that he embodies a particular clash of man against world that's both intoxicating and scary. He's aware this doesn't always endear him to people. When I ask him what drives him, he's disarmingly honest.

"Request for recognition. Ego." A pause. "Arrogance."

He smiles.

"You know what fighter pilots use for birth control?" he asks. "Their personalities."

My next stop is **XCOR Aerospace** for a meeting with CEO Jeff Greason. XCOR interests me because it's one of the few startups in this industry that isn't funded by a billionaire. You can tell that from their offices. Well, office (singular). Well, "office' is pushing it. Jeff and I meet in a shabby windowless hut furnished to give that authentic frontier vibe. Some of the seats are from old airplanes. There may be one of XCOR's bijou rocket engines on a shelf, but simpler technologies like the vacuum cleaner are clearly not welcome here. As we talk, it starts to rain, the downpour drumming loudly on the corrugated metal roof.

"Not everyone has picked up on the David and Goliath aspect of XCOR," says Jeff. "Richard Branson is pouring more money into Virgin Galactic in a month than we've spent in our history. We've got NASA doing billions and billions of dollars of its thing. We've got Elon Musk pouring all of his money into SpaceX. Amazon.com's Jeff Bezos is funding Blue Origin. . . ."

And yet, despite their underdog status, XCOR keeps being mentioned in the same breath as its better-funded rivals. Talking to Jeff, you don't get any sense of ego. He's trying to run a business—it just happens to be one that wants to make spaceplanes. That XCOR exists gives me more of a sense of a real spaceflight industry putting down roots. There's something very "regular Joe" about Jeff. He has the chunky mustachioed friendliness of your local car mechanic. Having served on the Augustine Commission, he can happily express informed views on national space policy but can instantly swap into a conversation about "propellant mix" or "fuel tank slosh." I also get the sense that Jeff has arrived at the place he's supposed to be.

"As an eight-year-old growing up in rural Oregon I found this book called *Door into Summer* by Robert Heinlein in the school library," he tells me. "The main character takes a military application, figures out how to make a marketable product out of it and starts a small company that sets up in a hangar in the Mojave desert . . ." He's got a smile on his face that's broader than the state he grew up in.

Jeff shows me around XCOR's hangar where the company is building the Lynx spaceplane—a two-seater craft no bigger than a navy fighter that will take off and land on a normal runway carrying one passenger or an experimental payload. I sit in a fiberglass mock-up of the Lynx's cockpit and it's at this moment that it really hits me. These people aren't joking. They really are going to open up a new industry. I look out of the cockpit and see engineers poking tools into rocket engines while Bon Jovi plays over the radio. Less than two hundred yards away, Scaled Composites is building *SpaceShipTwo* not as a one-off, but as an actual production spacecraft.

"The idea is that we always fly full," says Jeff. "It'll just be you and the pilot. We also have complete reusability, it's a 'gas and go' model. That's never been done. We're trying to make travel to and from space economically sane."

Jeff reckons his break-even point is just two hundred one-hour flights a year at ninety-five thousand dollars a pop. Given that the Lynx is a "gas and go" spacecraft, he estimates each one can fly four times in a working day.

In one corner of the hangar is XCOR's experimental EZ-Rocket aircraft, a precursor to the Lynx spaceplane that can climb into the sky at a dizzying ten thousand feet per minute.

"Dick Rutan was one of the test pilots," Jeff tells me.

"Interesting guy," I say. "He's out to disprove Einstein."

"Sounds like Dick." He laughs. "He's what you might call a strong personality . . ."

It's my last night in Mojave, and Jeff and I meet for dinner, an Ultimate Omelette at the local Denny's. "Ultimate" is something of a misnomer. I refuse to believe that I have enjoyed the pinnacle of omelette-kind. I particularly refuse Denny's the claim that it can deliver me that zenith. I am resolute. The quest for a better omelette will not die in Mojave.

Jeff picks up on a theme I've heard a lot in my time here: environmentalism. It'd be easy to characterize the risk-taking, military-minded, hunting, hard-living, swearing, frontier-loving personalities that I've been talking to as unthinking searchers of novelty who appropriate resources to satisfy their personal ambitions with little care for the ecological impact. But that would be wrong. While nearly everyone I've met is dubious about man-made global warming (Dick Rutan called it "a pathetic fraud, totally absurd and basically criminal") all of them are strong advocates of sustainability.

Stuart Witt had asked, "How would you explain our unsustainable use of fossil fuels for energy creation to an alien? It's rather barbaric. So doing this thing we now dub 'green' is literally doing the right thing. This isn't about politics, it's about doing the right thing as stewards of spaceship Earth."

"I can't support the industrial-scale reckless waste of the planet," says Jeff. "I like this planet. It's where all my stuff is. I think we should manage the planet better because, like it or not, we are in charge of it."

For the people here, it makes no sense to damage your own spaceship. Getting into space is a mental frontier that asks us to consider the Earth as a vessel that is carrying humanity. While most of us understand that idea in abstract, widespread access to space will bring the message home, they argue.

"With space as a single frontier, the entire planet is one civilization," says Jeff. "It transforms the mindset. That's the idea that keeps me going. That's the big one."

It's the complete opposite of the ambitions that characterized the space race. The spirit of Mojave has nothing to do with being "the biggest, baddest kid on the block." It's better summed up as "humanity, you need to get out more."

A lot of people might argue that we don't need to go into orbit to make us reflective about our role as stewards of the Earth, but as access to space becomes cheaper it can't help but subtly shift our perspective as a species. There isn't an astronaut on the planet that came back without some kind of transformation in their thinking.

My friend Bridget likes to say, "We're all environmentalists; we just haven't discovered it, and that's because we're all indigenous, but we've forgotten it."

It may be that opening up space to the masses could help jog the memory.

As my bus pulls out of Mojave I'm convinced that commercial space-flight is set to become a reality. Driving around the airfield with Stuart I'd seen real spaceships poking their noses through hangar doors.

"I don't think you'll see the first commercial operations out of this field until 2015 at the earliest and probably 2018," he'd said, which is a prediction some years behind those made by eager-beaver space fans and industry PR companies. Nonetheless, it's still an arresting thought. By the end of the decade the number of people who have taken a trip to space will, if you'll excuse the pun, start to rocket. It'll be a long time before the average Joe can take a trip though (not many of us have a hundred thousand dollars to spare for a couple of hours of sightseeing) but if commercial spaceflight goes the way of aviation, I may fulfill a childhood dream before I die. If I come back here to take that flight, the town will no doubt be transformed. By then that gleaming space port of the imagination could be a reality and my friend from Mike's Roadhouse Café and Bar will have to find a new place to live if he wants to avoid the temptations and trials of a thriving town.

Mojave could become a byword for glamour.

For now, however, my method of travel is slightly more mundane, although my destination is not. I'm heading to Washington, D.C., to meet one of the most important inventors in history.

8

"A Constant and Complete Intercourse"

> We should not only use the brains we have, but all that we
> can borrow.
>
> —*Woodrow Wilson*

I live on a hill, and at the top of that hill is a small park. It's part of the lifeblood of an area that, to be honest, is a long way from being a prestigious address. New Cross Gate SE14 is to Beverly Hills 90210 what The Troggs are to Bon Jovi (which is one of the reasons I like it).

On summer evenings the neighborhood tends to gather in the park to play Frisbee, drink, flirt, watch the sunset. From our position in the suburbs of the southeast corner of the city, we can see as far as Alexandra Palace ten miles to the north and Chelsea Harbor six miles to the east. The Houses of Parliament, the London Eye and Wembley Stadium are laid out for our pleasure. As dusk descends, the cityscape begins to twinkle as seven and half million souls reach for a light switch. It's beautiful. The same hilltop position that makes our sunsets so glorious also made it the ideal place, in 1795, to locate an "optical telegraph station," one link in an eighteenth-century information network connecting capital to coast.[1]

Optical telegraph stations were invented in France in the 1790s by brothers Claude and René Chappe, who were obsessed with finding a way of quickly communicating messages without using horses or pigeons. They started off by hitting cooking pots loudly at each other but found this method not only annoyed the neighbors (and anyone needing a saucepan) but wasn't effective over very long distances, especially if the wind was in the wrong direction. Instead they adopted a simple but effective visual system using a set of three connected wooden "arms" (two short arms jointed to a longer one between them) that could be placed in ninety-eight different configurations by using a system of pulleys housed in a hut underneath.

Providing there was a clear line of sight between huts, a message could quickly be communicated over long distances in short order (especially if the hut operators had telescopes). Messages that took days to travel by horseback could be transmitted in tens of minutes.

Napoleon Bonaparte was a big fan and, being convinced of success in a planned invasion of England, asked a third Chappe brother, Abraham, to design a telegraph that could signal across the Channel. He even built the French-side station at Boulogne.[2] Ironically it was the homegrown British telegraph system (which used a system of six large shutters rather than movable wooden arms) that brought the message of Napoleon's defeat at Waterloo in 1815 to London. This was twenty years after my neighborhood of Plow'd Garlic Hill became Telegraph Hill with the arrival of our own signaling hut.

The telegraph was heralded as a force for world peace. The 1797 edition of the *Encyclopaedia Britannica* suggested that "the capitals of distant nations might be united by chains of posts, and the settling of those disputes which at present take up months or years might be accomplished in as many hours."[3] As the relatively limited optical telegraph network gave way to the worldwide phenomenon of electric telegraph in the second half of the nineteenth century, the "force for peace" argument was being trumpeted again. In his brilliant history of the telegraph, *The Victorian Internet*, published in 1998, Tom Standage quotes a toast proposed by

British ambassador Edward Thornton upon the completion of the first viable transatlantic telegraph cable: "What can be more likely to effect [peace] than a constant and complete intercourse between all nations and individuals in the world?" asked Thornton, before going on to suggest that the telegraph would remove misunderstandings and promote "peace and harmony throughout the world." Each new communications technology is heralded as an opportunity for international friendship and understanding to flourish—and today's Internet is no different. Standage concluded that such hopes for our latest communications network are overstated. "That the telegraph was so widely seen as a panacea is perhaps understandable. The fact that we are still making the same mistake today is less so." But while the Internet (and its predecessors) have clearly *not* delivered world peace, are they *part* of the reason the world is getting less violent?

Yes, I know. I was surprised too. The world is getting *less* violent? The first time this was put to me, my reaction was, "Have you *seen* the news? Are you *crazy*?" But the figures speak for themselves and they speak volumes. It's an astonishing and underreported fact that violence is declining and has been for centuries. This goes against popular sentiment that the past was somehow safer and simpler—a time without nuclear weapons or helicopter gunships, with no violent movies or adults-only computer games. Steven Pinker sums up this misconception when he describes it as "the idea that humans are peaceable by nature and corrupted by modern institutions."[4]

In the last thirty-five years, anthropologists like Carol Ember and Lawrence Keeley have been scouring the archaeological record and studying tribal cultures with results that seriously question the idea that the trappings of civilization corrupt us toward violence. Their conclusion is startling. As Pinker writes: "the romantic theory gets it backward: Far from causing us to become more violent, something in modernity and its cultural institutions has made us nobler."

Rates of savagery (defined as the likelihood of being killed by another human being) can run to sixty percent in tribal societies and tend not to be less than twenty percent.[5] "If the wars of the twentieth century had killed the same proportion of the population that die in the wars of a typical tribal society, there would have been two billion deaths, not 100 million," writes Pinker. "Global violence has fallen steadily since the middle of the twentieth century."

I'm so (pleasantly) stunned by this thought that I decide I had better check the figures. I find myself searching the databases of the World Health Organization for causes of death, reading reports like the University of British Columbia's *Human Security Brief* (with chapters titled "Trends in Armed Conflict" and "Deadly Assaults on Civilians") and trawling through spreadsheets with file names like "Armed Conflict Dataset 1946–2008" offered by Uppsala University's Conflict Data Program. It doesn't sound like a bundle of laughs,* but it's surprisingly heartwarming, if you can get yourself above the fact that every statistic represents the culmination of a fellow human being's hopes and fears wiped out by a bullet, bayonet, land mine or cluster bomb.

Just 2.8 percent of the world's population died as result of violence in 2002 according to the World Health Organization. Over half of those deaths were the result of suicide (self-violence), a further one percent homicide and just 0.3 percent due to conflict.[6] The WHO figures for the previous three years show a similar pattern.

Reading *Human Security Brief* you discover the number of annual battle deaths in interstate wars fell from over sixty-five thousand in the 1950s to below two thousand per year in the first decade of the twenty-first century. The document is full of not-in-the-news statements like "from the beginning of 2002 to the end of 2005, the number of armed conflicts being waged around the world shrank 15%" and "notwithstanding the recent increase in terrorist attacks, the number of civilian victims of intentional organized violence remains appreciably lower today than it

---

*I get *déjà vu*—after all, I started this trip researching life expectancy figures.

was in the Cold War years." There's more: the 1990s, it says, was the "first time more wars (42) ended by negotiated settlement than by military victory (23). This started a trend that accelerated in the new millennium. Between 2000 and 2005, 17 conflicts ended in negotiated settlements; just 4 ended in victory." The authors conclude, "On average over the past six years, more conflicts have stopped than started each year. There is no reason to expect this trend to continue, but nor is there any reason to expect it to be reversed."[7]

There's a problem with the figures I've uncovered, though. First, they're only a snapshot covering a short period of recent history. Second, statistics about levels of violence are extremely patchy. There is no reliable and consistent source of data for war deaths (nations in the grip of conflict tend to make bad record keepers). Third, all such figures are highly controversial. The methods by which data is collected and reported (and the political motivations for doing so) are regularly challenged. Recent research by the Institute for Health Metrics and Evaluation in Washington and Harvard Medical School suggests that in some cases war deaths may be *three times more* than WHO estimates.[8] Currently the UK-based Oxford Research Group (a charity "which works together with others to promote a more sustainable approach to security"[9]) is trying to implement a worldwide framework for counting the dead.[10] As yet it doesn't exist.

All that said, the long view is that the underlying trend in violent deaths is steeply downward. Even if we double, triple or quadruple the recorded rates of slaughter for the last century we're still killing far fewer people per capita than our ancestors. As Steven Pinker says, "We must have been doing something right. And it would be nice to know what, exactly, it is."

My hunch is that one of the things we're getting right (among many others) is becoming increasingly connected, and being so we find it harder to kill each other. It's a popular view and one that seems to make instinctive sense. I'm conscious though that belief and truth should never be confused (as anyone who's attempted an LSD-induced flight off the top of a car park might, if they survive, tell you). I call some notable Peace Studies academics to see if there is any research that proves or debunks the theory. The Department of Peace Studies at the University of Bradford tells me my question is "flimsy" and "impossible to prove." The widely respected history professor and author of *An Intimate History of Killing*, Joanna Bourke of Birkbeck University, replies to my e-mail with "I feel really stupid, but actually have no idea."

The most evidence I can find are articles that give specific examples of communications technologies which help antagonistic parties share and process information, talk to each other or improve decision making in a nonviolent direction. So, my hunch remains just that, backed up by a few scattered examples and a simple and seductive logic: that talk, trade and the sharing of culture across boundaries joins our fates together, giving us less incentive to kill each other.

This is what American philosopher Robert Wright refers to as the "nonzero-sum game," by which he means "just because I win doesn't mean you have to lose." My local park can be seen as a nonzero-sum game. It's not just a nice outdoor space: for families without a garden it's a place for their children to run free, and for the rest of us it's a space we can share without fighting for it. The restoration of the park in 2004 was a local effort spearheaded by Malcolm and Jayne, whose win was a win for all of us. This "nonzero-sum" ethos is a strong part of my local community. John, Catherine, Patricio and Stephen are spearheading the construction of a much needed local café, Bridget runs the summer festival, and Jules, Sara, Susan, Martin, Phil, Max and Rosie give us an arts festival every Easter. Your body is nonzero-sum game too—each cell benefits from every other: if your liver wins, the rest of you tends to as well. (The converse is also true, as you'll know if you've ever had a hangover.)

Wright's argument is that "through technological evolution new forms of technology arise that facilitate or encourage the playing of nonzero-sum games involving more people over larger territory."[11] This technology includes agriculture, urbanization and, of course, communications and manufacturing technologies. Technology helps us create a worldwide park and café. To make the point he says, "If you ask me why am I not in favor of bombing Japan I'm only half joking when I say, 'Well, it's because they built my car.'" (It's worth remembering that sixty-five years ago America dropped two atomic bombs on the Japanese—an action which would seem inconceivable today.) Slowly we're playing a bigger "nonzero-sum game" with more and more of us involved.

Before you get the idea that Wright isn't acknowledging the troubles we face, he states his theory is "not intrinsically upbeat, it can accommodate the existence of inequality, exploitation and war. . . . all it tells you is the fortunes will be correlated for better or worse. It doesn't necessarily predict a win-win outcome." Taken over all of history, he argues that "people have played their games to more win-win outcomes rather than lose-lose outcomes. On balance I think history is a net positive." He argues that there is a growing moral sense in the world, that the majority of us now believe that "all people everywhere are human beings and deserve to be treated decently regardless of race or religion." While the dehumanizing effect of ethnic or ideologically based conflict is still all too obvious, it is no longer, says Wright, the prevailing view if you take the world as a whole. "You have to read your ancient history to realize what a revolution that has been. This was not a prevalent view a few thousand years ago."

It's a desire to talk about our increasing interconnectedness and how this will impact on our future that finds me approaching a large house in McLean, Virginia, just outside Washington, D.C. We keep hearing about the Networked Age, but is it a good thing? Is it one of the reasons we're fighting less? Perhaps the man I have come to see can bolster or crush my belief.

This is by far the most affluent neighborhood I have ever set foot in. Mansion-sized houses sit amid immaculately kept lawns. Wide driveways accommodate multiple expensive and impossibly clean cars gleaming in the sunshine. (I can't prove it, but I suspect this is the sort of neighborhood that has more fridges per household than children.) You don't get to live here unless you've succeeded at something.

I'm still a good ten yards from the door when it opens in anticipation to reveal a tall, healthy-looking man in his mid-sixties, immaculate in a three-piece suit, with a welcoming smile framed by a well-kept gray beard. "Hello," he says and extends a hand. This is Vinton, known as "Vint," whose success has had a fundamental impact on the world. Vint is one of the men who *invented* the Internet.

As we take seats in the library (I told you it was that kind of neighborhood) he hands me his card: "Vint Cerf. Google. Chief Internet Evangelist. I tried for archduke," he jokes. The job title, he tells me, is a reflection of the fact that he's "spent decades trying to get more Internet built and persuading people it's a good thing to do." It's a mission he's still on. "It's something I still need to do because there's only one-third of the world's population online," he says. Vint believes in the virtues of interconnectedness.

I give Vint my card. "This is sort of silly," he says, taking it. "It's the twenty-first century and we're handing out little pieces of cardboard." As our conversation progresses I'll learn that Vint has an eye for the silly. A recent invitation to a gathering of Catholics saw him arrive in traditional Spanish academic regalia (a costume given to him at one of many ceremonies attended to receive his eighteen honorary degrees). When asked what religion he represented, Vint replied, "Geek Orthodox." He wore the same outfit on his first day at Google. "I didn't want the young Googlers to think I was some boring old fart, and if they did, I at least wanted them to think I was an *unconventional* boring old fart," he chuckles. At all other times though you'll find Vint dressed in a sharp three-piece suit. "It's an ingrained habit of not wanting to look like everybody else," he tells me. "I wear them all the time. I wear them on flights so if I lose my luggage I'm still properly dressed for any event. I remember having to meet the Bulgarian president and prime minister and the airline lost my luggage for four days. I got along just fine." Vint Cerf: old school, meticulous, playful—and smart as hell.

I'm hoping Vint can give me the big picture on our increasing interconnectedness. After all, he was in at the ground floor of the Internet and now works on the top one. He's a man with a career-length view on the technology, which for a technology as young as the 'Net is about the longest view you can have. As a graduate student, Cerf worked under Professor Leonard Kleinrock, who in 1969 oversaw the first computer-to-computer message to be sent using the "packet switching" method that underlies the Internet. Actually, it was two-thirds of a message. Another of Kleinrock's students, Charley Kline, hoped to send a three letter message "LOG" to a receiving machine (this being the code for logging on to that computer). The "L" and the "O" worked but the "G" crashed the system. "So the first message on the Internet was LO," said Professor Kleinrock. "Or 'Hello,' crash!"[12] Cerf and others went on to help create ARPANET, a U.S. government research project led by Lawrence Roberts. The ARPANET is often mistaken for being the first incarnation of the Internet we know today, but Vint tells me, "The ARPANET became one of several networks that were interconnected to become part of the fledgling Internet."

Vint is regularly dubbed "the Father of the Internet," an accolade he says is "just wrong" and the result of "an overactive public relations group" at his former employers, MCI. "It causes a lot of trouble, because there are a lot of people that deserve credit. It's really painful for me personally to hear people say that and to have to try and defend against it," he says. But if Vint isn't *the* father of the Internet he's one of them. It was Vint and his colleagues (notably Bob Kahn) who answered the question "How can we reliably get data that lives on a computer on *this* network to another computer on *that* network when both might be using different hardware and software?"—a problem then called the "Inter-Net problem." Their answer was a couple of "software protocols" with the decidedly unsnappy title of "Transmission Control Protocol/Internet Protocol" or TCP/IP for short. (You may have come across this un-friendly couplet when you've been battling with your Internet provider.) On the Internet there is no fixed circuit set up between origin and desti-nation along which data can pass. Instead, data is split up into numbered "packets" which are sent into the wilds of the 'Net to find their own way to the destination. As each packet passes through any of the routing points on the network, it says, "hey, I'm trying to get to B, do you know where B is?" and one of three answers will come back: "Yes! *I* am B," "Yes, B is over there" or "No, but I'm sending you to another machine who might know where B is." When I check this summary with Vint, he tells me, "Well, it is a *bit* more organized than that! But you are not far off!"

In a 2009 TED talk, Jonathan Zittrain, professor of Internet law at Harvard Law School, explained this process by asking his audience to imagine they were at a sporting event and somebody asks for a beer. "It gets handed at the aisle and your neighborly duty is to pass the beer along, at risk to your own trousers, to get it to the destination. No one pays you to do this. It's just part of your neighborly duty. That's exactly how packets move around the Internet, sometimes in as many as twenty-five or thirty hops with the intervening entities passing the data around having no particular contractual or legal obligation to the original sender or the receiver." Vint nuances this description. "Actually most connections *do* involve contracts, even if no money changes hands, but the arrangements are entirely voluntary."

At its heart the Internet is a system that works on trust and cooperation. It is itself an example of a nonzero-sum game. All systems cooperate, so all systems can benefit from the network. The software freely collaborates to split up, route and reassemble those packets of data so you can e-mail your friends, store and retrieve files from the office, watch movies, video conference with your relatives in distant lands, and see Paris Hilton in all her nighttime glory. "If you look at the network itself, it is a vast collaboration, there is no central authority, and nobody is forced to build and operate pieces of it," Vint says. "I think that it should be self-evident that this notion of cooperation and mutual self-interest is an extremely powerful part of the Internet story."

At this juncture it's worth clearing up a common point of confusion— "the Internet" and "the Web" are not the same thing, despite often being referred to interchangeably. The Internet is deep plumbing. You won't really see it, any more than you see the sewer when you go to the bathroom. The Web, invented in 1989 by Tim Berners-Lee, sits on top of that plumbing to provide us with a service—a way to take packets of data and present them in a visual and interlinked way called "web pages." (The Web is not the only service sitting on top of the Internet; e-mail is another.) The two terms, however, have become popularly interchangeable because it was the Web that suddenly made the Internet useful to a much larger audience.* While the implications of mixing the terms up are not as bad as confusing "sewer" for "bathtub" when suggesting the location for a romantic moment, some geeks do tend to get niggled by this kind of thing. My friend Paul (a website designer) gets around this by calling the combined experience the "infowebulator." Vint himself says there's nothing wrong with confusing the two "with the exception that it could potentially limit your thinking about what you can do with it."

Beyond personal uses of the Internet/Web combo (like being able to shop, get travel information or find pictures of humorously ugly cats†) there are collective benefits too. "One thing I will observe about the Internet which I find to be different from television, radio, the telegraph, telephone, newspapers and magazines is that it permits a kind of group interaction that has never been possible before," says Vint. "A larger number of people have the option of interacting with you—and I think that's important."

The telephone (and its telegraph predecessors) allow one-to-one conversations, but those conversations can't cope well with large numbers of people (as anyone who's been in a conference call will tell you). Videoconferencing suffers from the same constraints. Television, radio and newspapers are the exact opposite, reaching large numbers, but there's little dialogue—they broadcast, we receive. The Internet/Web provides a

---

*When Berners-Lee wrote a memo suggesting a "distributed hypertext system" to his boss, it was returned with the words "vague but exciting" written on the front page. "Eighteen months later my boss said I could do it on the side as a sort of a play project," he recalls. That "hypertext system" became the Web.

†Why not check out www.ugly-cat.com? Or better still: www.catsthatlooklikehitler.com.

platform to bridge that gap. Individuals across the world are able to form groups much more quickly and powerfully than at any time in history. Toenail collectors can now find other toenail collectors (I'm not kidding). Protest movements can achieve the critical mass needed to have their voice heard in a way that simply wasn't possible in a pre-Internet age. (A small example is the online petition that convinced the British government to issue its posthumous apology to computer science war hero Alan Turing. A large one is the Iraq war demonstrations that used the power of the Internet to mobilize millions of demonstrators across the globe.) In between are groups of enthusiasts who collectively craft Wikipedia entries for the benefit of us all, or develop "open source" software tools, or form online communities to discuss everything from death metal to Marxist politics. My favorite example is The Legion of Extraordinary Dancers and I'm not even into dance. Ten years ago this group could not have existed. In 2009, they performed at the Oscars and are now regarded as a dance phenomenon.

They exist because kids dancing on street corners started filming themselves and putting their moves on YouTube. Kids from Detroit started innovating on moves created by dancers in Tokyo, and within hours new dance moves would evolve. The best dancers of the moment, with the craziest moves, created what is known as a "spike"—a beacon that drew others to them, an audience they could never have reached before. Social networking sites buzzed with friends linking to videos of cool moves. *Look at this guy!* John Chu, who created The Legion of Extraordinary Dancers from finding the most popular of those YouTube clips, says, "Dance has never had a better friend than technology. Online videos and social networking have created a whole global laboratory online for dance."[13]

Marketing guru Seth Godin would call The League of Extraordinary Dancers a tribe:

What tribes are, is a very simple concept that goes back fifty thousand years. It's about leading and connecting people and ideas. And it's something that people have wanted forever. Lots of people are used to having a spiritual tribe, or a church tribe, having a work tribe, having a community tribe. But now, thanks to the Internet, thanks to the explosion of mass media, thanks to a lot of other things that are bubbling through our society around the world, tribes are everywhere.[14]

The argument is that the Internet helps us find our tribes, or take charge of a tribe that's been waiting to be led. ("The Beatles did not invent teenagers," says Godin. "They merely decided to lead them.") The Internet then is the great equalizer. You no longer need huge amounts of money or a broadcast network, you just need an idea and a leader whom enough people can gather around. It's not that the Internet suddenly makes us a global family, but it does allow us to create "families" that cross boundaries of geography and wealth with greater ease, so many argue this can only be good for fostering a brotherhood of man. Beyond this the Net stimulates innovation, not just in dance moves but in science, technology, political thought and ugly-cat theory. The Internet becomes the crucial engine that powers the information age—a "Knowledge Combustion Engine" if you will.

But there's a dark side too. Increasing connectedness helps violent ideologues, pedophiles and bullies find each other (and their victims) too. Another problem is that prejudice can be fostered rather than dissolved. In 2010, John Seely Brown, one of America's most respected thinkers on the interplay of technology and society, said, "The blogosphere and the echo chambers you find there are turning more groups violent. They're amplifying the extreme. You tune into the echo chamber you like best. There's no particular incentive to listen to ideas you don't like. We're going to have serious terrorism, self-generated in this country very soon, because of the Internet."[15] Gabriel Weimann, professor of communication at Haifa University, argues that "by its very nature, the Internet is in many ways an ideal arena for activity by terrorist organizations."[16] The Internet offers ease of access, can reach large numbers of people, operates under almost no regulation or censorship, is cheap to use and can provide anonymity. In short, it's the ideal propaganda engine for the evildoer.

Robert Wright adds another angle to the "Internet as terrorist incubator" argument: "More and more, it's possible for grassroots hatred abroad to manifest itself in the form of organized violence on American soil. And that's pretty new, and I think it's probably going to get a lot worse—this capacity—because of trends in information technology [and] in technologies that can be used for purposes of munitions—like biotechnology and nanotechnology." Those nanotech and biotech 9/11s I'd discussed with George Church and Eric Drexler will be aided and encouraged into reality by Internet-enabled groups of extremists.

It's the age-old battle of freedom of speech versus the proliferation of extremism. "In the United States we pay a very notable price for our freedom of speech," says Cerf. "It's hard for us sometimes to tolerate people saying things we don't agree with. But to preserve the society we live in you have to fight tooth and nail for that person's privilege to say what they just said. Not everyone in the world feels the same way about that." The Internet exposes striking differences between nations. "The European view of privacy and freedom of speech is different from the American version. But because the Internet is global and largely nonnational in character it lands in the middle of all of these different, conflicting sociological perspectives." Soon after my visit to Washington, Pakistan blocks access to Facebook and YouTube "in view of growing sacrilegious content," demonstrating the tensions between politics, religion and free expression.[17] (Talking of sacrilegious content, Facebook has recently revealed that an old school friend of mine, the previously aloof Tessa, is now a pole-dancing fanatic.)*

Reporters Without Borders, a nonprofit organization based in France that campaigns for press freedom, maintains a list of "Internet Enemies"—as I write, Burma, China, Cuba, Egypt, Iran, North Korea, Saudi Arabia, Syria, Tunisia, Turkmenistan, Uzbekistan and Vietnam.

---

*Oh to have a time machine.

On its "surveillance list" is Australia, which has proposed legislation to implement what Reporters Without Borders calls "a draconian filtering system."[18] Vint is concerned too: "Australia is apparently making an attempt right now to filter the 'Net, under the guise of protecting children." The *Brisbane Times* reported that the Australian blacklist of sites at one point included "a slew of online poker sites, YouTube links, regular gay and straight porn sites, Wikipedia entries, euthanasia sites, websites of fringe religions such as satanic sites, fetish sites, Christian sites, the website of a tour operator and even a Queensland dentist."[19] (I checked the list on Wikileaks.org and it does indeed include dentaldistinction .com.au.[20]) While I'm sure many Australian children might actually *welcome* the government trying to keep them away from dentists this does seem a bit over the top, as does blocking the sites of "Maroochy Boarding Kennels" and "Tuckshop and Canteen Management Consultants." It's hard to imagine what the Australian government thinks its citizens have to fear from these organizations, unless there's a conspiracy afoot. Are boarding kennels selling dog meat to tuckshops? Is it dog meat that accelerates tooth decay? I think we should be told.

As Vint and I sit talking, Google is battling with Beijing over the censorship of Web searches, a tussle that will result in the company shutting down its China-based search engine. It's another example of how interconnectedness doesn't always bring us together. The overall picture is "a brow-wrinkling experience," says Vint. "I wish I was eight years old so I could see how it all plays out in the next fifty years."

Another problem with increased interconnectedness is that while violence is declining, we fear it more. Vint attributes this to the fact that today we "experience more of the world than we would otherwise, and sooner too—almost in real time." This means "we can misunderstand or misapprehend what it is that we have just learned or discovered or encountered." Instant and sometimes unfiltered reporting of violence means "you start to get the feeling that you are at risk, that the world is a dangerous place," says Vint. "Perspective starts to leak away because of this connectivity." To make the point he refers to riots in Kyrgyzstan that are going on as we speak. In a less connected world, such violence in far-flung places wouldn't have even registered with us. Now, we hear about it wherever it is, and instantly. (The upside is that news of violence can "get out" quickly. YouTube and wireless connections make brutality much easier to shine a light on.) The result can be that even as violence declines, we fear it more and, being fearful, we feel the need to defend ourselves against a dangerous world. Robert Wright has suggested this could be called "the positive feedback of negative vibes" by which he means that negativity feeds negativity, not that there's anything "positive" (in the conversational meaning of the word) about it. "We really shouldn't have the word 'positive' there at all, even in a technical sense," he says. "So let's call it the death spiral of negativity." Oh yes, let's.

The idea that interconnectedness makes us less likely to fight is starting to take a few knocks, but none of them strike me as fatal blows. Certainly interconnectedness increases the instances of some kinds of violence, but for every "death spiral of negativity" it's easy to find a League of Extraordinary Dancers, or a worldchanging.com (the influential hub that champions "innovative solutions to the planet's problems") or a couch-surfing.org (a website used by millions that hooks up backpackers looking to sleep on a stranger's couch with people who are happy to have random travelers kip on their furniture). If you look at the numbers, people use the Internet to connect with each other and learn. The world's most popular sites besides search engines are social networks, blogging tools and online encyclopedias.[21] On top of that, the myth that the Internet is dominated by pornography also turns out to be just that. In 2006, Philip Stark at the Department of Statistics at the University of California, Berkeley, was commissioned by the U.S. Department of Justice to look into the effectiveness of Internet filtering software and has been called to testify as an expert witness in censorship cases. He concludes, "About one percent of the websites in the Google and MSN indexes are sexually explicit. About six percent of queries retrieve a sexually explicit website."[22] That said, "Nearly forty percent of the most popular queries retrieve a sexually explicit website," although Philip tells me that just because a search term may return a link to naughtiness that doesn't mean "the person who ran the search was looking for it or visited it." After all, lots of terms have double meanings, most obviously "sex," which can refer to gender as well as horizontal athletics. The widely held belief that "the Internet is awash with pornography" is perhaps better restated as "the Internet is about one percent pornography, but that one percent is pretty popular."

It's nearly six hundred years since the invention of the printing press, over a hundred and fifty since the invention of the electric telegraph, more than one hundred and thirty since the telephone, over eighty since TV and twenty years since the Web made the Internet a useful tool for everyday folks. Yet because there is no research that can tease out the links between rates of savagery and levels of interconnectedness, my hunch that it's one of the reasons we're fighting less remains just that. Still, the last time I checked, the Web appears to have more pictures of people making love than war and so I'm sticking with my gut feeling.

Back in Cerf's library our conversation turns to more upbeat territory. "My optimistic statement of the day is not that information is power, but that information *sharing* is power," he says. "I think that's repeatedly demonstrated in the course of human history—that the sharing of information makes us all more powerful—and that any society that suppresses information harms itself in large measure." As Stephen Hawking famously said, "With the technology at our disposal, the possibilities are unbounded. All we need to do is make sure we keep talking."[23]

The Internet is another forum in which to talk. Like the telegraph, the telephone or the newspaper, it's not perfect, but it adds to the number of ways we can converse to share information and ideas. "The good side of it is that we encounter people we never would have encountered, we have an opportunity to rub ideas together we might never have had the chance to explore—and I think that's incredibly powerful," says Vint.

Perhaps this is why staff at the Italian edition of *Wired* magazine nominated the Internet/Web for the 2010 Nobel Peace Prize and Vint Cerf, Bob Kahn, and Tim Berners-Lee to be the recipients if it's accepted.

But connecting *people* is only a fraction of the Internet's story.

Cerf has written:

In the next decade, around 70% of the human population will have fixed or mobile access to the Internet at increasingly high speeds. We can reliably expect that mobile devices will become a major component of the Internet, as will appliances and sensors of all kinds. Many of the things on the Internet, whether mobile or fixed, will know where they are, both geographically and logically. As you enter a hotel room, your mobile will be told its precise location including room number. When you turn your laptop on, it will learn this information as well—either from the mobile or from the room itself. It will be normal for devices, when activated, to discover what other devices are in the neighborhood, so your mobile will discover that it has a high-resolution display available in what was once called a television set. If you wish, your mobile will remember where you have been and will keep track of . . . objects such as your briefcase, car keys and glasses. "Where are my glasses?" you will ask. "You were last within . . . reach of them while in the living room," your mobile or laptop will say.[24]

"There's the 'Internet of things' in addition to the Internet of people and ideas," says Vint. As computing technology continues to get smaller, almost every object has the potential to become a node on the Internet. It's a world where if you've lost your keys you'll Google them, where your fire alarm will call you to let you know it's been activated and your toothpaste orders more of itself as you reach the end of the tube.

"I think that 'everything is connected' will not only be a technically correct statement in the mechanical world as, over time, more and more things literally are connected, but it's *always been true* in the conceptual sense," says Vint.

It turns out Vint and I are both fans of James Burke, the British science historian and one of the faces of popular science on television in the seventies and eighties. Mine and Vint's favorite show was *Connections*, where Burke teased out links between society and technology. "He's moderated some sessions at Google," Vint tells me. "He starts out with some fact and then he wanders around in 'connectivity space' taking you far afield and finally completes the loop back to the original point. I had a good deal of fun with him."

Early in the first episode of *Connections*, first broadcast in 1978, Burke says, "The story of the events and the people who over centuries came together to bring us in from the cold and to wrap us in a warm blanket of technology is a matter of vital importance. Since more and more of that technology infiltrates every aspect of our lives it's become a life support system without which we can't survive."[25] Is the Internet the latest part of that ever growing support system? Certainly it's hard to imagine how many economies would cope if it suddenly stopped working. Physicist and computer scientist W. Daniel Hillis has written, "We are now all connected, humans and machines. Welcome to the dawn of the Entanglement."[26]

I put this to Vint. "This is not new," he says. "We have always been entangled with our technology, we've always been entangled with knowledge. It may be more obvious now, because of the way it manifests. But if you were a cave man you might have become quite dependent on tools that you built, because without them you might not be able to feed yourself, so you needed the knowledge to make those, or you needed the knowledge to find somebody who could make them. And then you also had to know that *that thing* over there was a saber-tooth tiger and it was a really good idea to get away from it. The people who didn't understand that didn't survive." In short, entanglement with knowledge and technology keeps more of us alive. The more connections the better.

But there's another worry. If the Internet is now, as Burke foreshadowed, a technology that "infiltrates every aspect of our lives" and "a life support system without which we can't survive," what happens if someone switches it off?

In 1909 E. M. Forster wrote a short story called "The Machine Stops." In it, a worldwide society entirely dependent and cocooned inside an omnipresent machine meets its end when the machine breaks down. Millions die and the human race is left to a handful of forgotten humans who still live on the surface of the Earth. Will they restart the machine? "Never," says the main character shortly before he joins "the nations of the dead." "Humanity has learned its lesson."[27] Let's be honest, this is not a cheery bedtime story for the kids. But it does play to one of our deepest fears.

Cerf is a big fan of the story. Could the Internet stop? Well, not easily. "Parts of it become inaccessible, parts are under attack, information disappears because somebody shut down the website, or disappears because we don't know how to interpret the bits anymore, pieces of it peel away and disappear," he says. "But I don't think you could stop the whole thing very easily."

The reason for this is that the Internet isn't *a* machine, it's *billions*. There is no "off" switch. Its very architecture, its decentralized, nonzero-sum collaborative fabric means that the Internet is a bit like the planet's population. Short of an apocalypse, you couldn't kill all of it, any more than you could wipe out everyone on Earth. And if there is an apocalypse, I'm guessing not many of us will be worrying about access to Facebook.

Kevin Kelly, one of the founders of *Wired* magazine, believes the Internet/Web will evolve into "the largest, most complex and most dependable machine we have ever built. It will also be the platform that most business and culture will run on"—something he calls "The One Machine" (even though it is in fact billions of computers connected together).

"Is there *anything* that could shut down the Internet?" I ask.

Vint considers. "If every internet service provider in the world decided one day just to shut down the routers, that would pretty much screw the Internet," he says. "So the answer is, it's *technically* possible but it would require cooperative action that's extremely unlikely." There is no conclusive count of the number of internet service providers on the planet but the figure is likely to run into tens of thousands. "So," asks Vint, "now the question is, is there *hostile* action that could shut the 'Net down? Well, there *are* hostile actions going on every day all the time and they're capable of rendering *parts* of the 'Net inoperable, but I don't think the machine would stop in and of itself.

"We launched the Internet in January 1983, and as far as I'm aware the entire system has never been shut down since." Like it or not, the Internet isn't going anywhere.

Technology's story is our story. Burke's "warm blanket of technology" isn't separate from us, we're woven into the fabric of it and vice versa. Imagining that technology is somehow not natural is about as useful as trying to separate your mind from your brain. In the next chapter of the Internet's story, intertwined with "the Internet of things" is something called "augmented reality," one of those phrases that strikes the same fear into my heart as those thin yellow burger slices that are "cheese flavored" and not *actual cheese*. What's wrong with *real* reality then?

A man walks into a shop and picks up a packet of paper towels. As he does so, an image appears on the packet telling him how much bleach was used in its manufacture. He picks up another and compares. The second gets a "green light" that appears as a ghostly image on the back of the packet, signifying eco-friendliness. He chooses the latter as his purchase. Now he makes a phone call, holding out his hand where the numbered buttons of a keypad appear sketched in light on his palm. He dials the number by tapping his own skin. Later, on the way to the airport, he pulls out his boarding pass, and across the top some text appears telling him his flight is twenty minutes delayed.

This sounds like a scene from a (admittedly quite dull) science fiction movie, but it is not. These are scenes from a demonstration of a new device developed at MIT's Fluid Interfaces Group—a combination of mobile phone, wearable camera and tiny projector that the lab's director, Pattie Maes, calls "SixthSense," a technology designed to provide seamless and easy access to "information that may exist somewhere that may be relevant, to help us make the right decision about whatever it is we're coming across," to help us "make optimal decisions about what to do next and what actions to take."[28] Vint thinks it's great. "It's a good example of what happens when your computer participates in the same real world that you do," he says, smiling.

Technologies like these start from the assumption that every object is surrounded by a "virtual" cloud of data you can't see. A piece of clothing isn't just the physical garment. In the shadow-world of data it is also how much it costs, whether it was manufactured ethically, the instructions for how best to wash it and so on. Crucially, your behavior toward it may alter with access to any one of these pieces of data. Imagine walking into a sneaker store and being able to instantly see, by looking at a product's bar code, whether it was made in a sweatshop or not, or if the shop around the corner had the same shoes on sale.

In 2010, Microsoft employee Blaise Agüera y Arcas demonstrated the ability to link up online maps with photographs and video, allowing you to take not only a virtual walk around an area (including "walking" *inside* buildings) but also to see what's going on at that moment, with real-time videolinks embedded into the scene. The same technology can be used to link historical photos and videos into the map, allowing you to step back in time. You can look into the sky and see star maps, or find out what blog entries refer to a particular place.[29]

While these two examples are at the cutting edge of "augmented reality," the layering data on top of our day-to-day experiences is already with us. Download the "Better World Shopper" app onto your iPhone, for instance, and it will give you an instant rating of a manufacturer's record in regard to human rights, environmental policy, animal rights, social justice and community involvement. "Google Goggles" makes use of your mobile phone's camera to recognize landmarks, book covers, even wine labels, and return Internet searches that relate to what you're pointing it at. This is data layered over reality, or (depending how you look at it) reality revealed by data. So, nothing like fake cheese after all.

"This ability to aggregate large amounts of information from many different sources in a coherent way has dramatically changed our ability to understand our world around us and to be aware of it and react to what's happening," says Vint. "Although I'm pretty sure that you could make a credible argument that these capabilities were present and have been present for a very long time in other media, they haven't been there in the same quantity, they haven't been there with the same immediacy and they have not been there to the same degree."

"We are Entangled?" I ask, referencing Daniel Hillis's quote.

"Yeah," says Vint. "We are Entangled."

A connected world is a good thing then, but it hasn't escaped most people's notice that we're being told the planet is facing a crisis that overshadows all others. If the worst predictions of climate change come true, there won't be many of us left to connect, and despite all the civilizing effects of connectivity, violence will rocket as wars over land and resources explode across the planet.

James Lovelock, the scientist who originated the famous Gaia hypothesis of the world as an integrated and carefully balanced system, has said that if it warms up as much as he thinks it might in the next hundred years, "We'll be lucky if there's a billion [people] left" (that means a cull of roughly eighty-five percent of all humanity).[30] Others suggest the warnings of climate change are a grand hoax or a scientific orthodoxy that's become too powerful to argue against.

This is *real* reality and something I can't leave out of my investigations, and so I'll soon leave America to visit the president of a nation that will entirely disappear if sea levels continue to rise and global warming makes good on the threat we're being warned about. But before I do I'm heading to the Big Apple to meet the American oceanographer who coined the phrase "global warming" and the German physicist who he thinks has a large part of the solution.

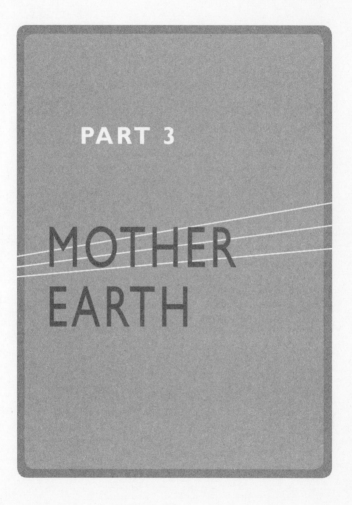

PART 3

MOTHER
EARTH

9

"World Leaders Still Don't Really Get It"

> A change in the weather is sufficient to re-create the world and ourselves.
>
> —*Marcel Proust*

Climate change. Go on then, *you* start this chapter. Because I don't know how to. I mean, do I start from the viewpoint that man-made global warming is real and then deal with arguments against it? Or do I start from a skeptical position and go the other way? And which set of readers do I put off in the process? Depending on your position you've probably had a sharp reaction to the previous sentences already. "Why would you even *consider* acknowledging arguments *against* global warming? Are you an idiot?" Or, conversely: "Oh you're one of those who's been suckered into this whole climate change scare-fest, are you? Moron."

I've tentatively raised the topic of climate change in numerous bars across America. The reactions have been fascinating. Some people launch into a polemic about how it's all made up, while others do the exact opposite, angry that their fellow countryfolk are blind to risks that seem obvious. A small minority shrug their shoulders and say, "I don't know what to think."

Climate change is such an *emotive* topic. It's one of those subjects, like religion and politics, that you tend not to raise at parties. In fact, it sometimes feels like religion and politics rolled into one. If you don't believe in the global warming threat, you could be branded a "climate denier," language borrowed from the world of Holocaust denial ("Well, it *will* be a holocaust if we don't do something about it," you might be told). On the other hand, advocating cutting carbon emissions may see you accused of falling for the greatest fraud in scientific history, unable to challenge an orthodoxy that's become a dangerous *fait accompli*, victim to hysteria overplayed by both scientists seeking funding and irresponsible media organizations who love nothing more than a catastrophe to get their teeth into. Both sides question the science and motivations of the other and suggest that their opponents live in some kind of intellectual echo chamber which amplifies their erroneous beliefs.

One thing everyone can agree on is that carbon dioxide levels in the atmosphere *are* rising. Ever since a man called Charles David Keeling started taking measurements of atmospheric carbon dioxide at the Mauna Loa observatory in Hawaii in 1958, the concentration of CO_2 in our skies has risen by about one quarter, from just over three hundred and fifteen parts per million in March 1958 to a notch over three hundred and ninety-two as I write in May 2010.[1] Ice cores containing trapped bubbles of air from yesteryear allow us to estimate how much CO_2 was in the atmosphere much further back, and the record shows a sharp incline since the Industrial Revolution onward. Another thing pretty much everyone agrees on is that global temperature has, on average, gone up by about 0.8 degrees centigrade since 1880.[2] There is also no argument over the fact that the amount of CO_2 produced by man-made activity is rising. After all, our fossil fuels are largely made of carbon (coal being the most obvious example), so when we burn them to release energy, much of this carbon finds its way into the sky (indeed it's inevitable unless we go to great lengths to prevent it). Finally, historical records over much longer periods show carbon dioxide levels and temperature are generally in step with each other (though not always).[3] So far it looks like a pretty conclusive case for the "man-made global warming" then, doesn't it? Surely we can't expect to put a trillion tons of carbon dioxide into the atmosphere without something happening? If you're convinced of the climate change threat, you're probably nodding your head, thinking "good start" and waiting for me to bring in the big guns of massive scientific consensus, disappearing polar ice, and more than a century of laboratory experiments that show how carbon dioxide absorbs heat radiation.

But I want to take a step back and ask why is it, for instance, that a national survey conducted in 2009 by the Pew Research Center (a deliberately nonpartisan organization funded by a charitable trust) found that thirty-three percent of Americans don't think there is solid evidence for global warming and a further ten percent can't decide?[4] Or why is it that climate change skepticism is in political ascendancy in Australia, where public opinion shows a similar pattern?[5] In the UK a number of recent polls suggest that anxiousness about climate change has dropped (although the majority of citizens are still concerned) while the number of "climate change agnostics" has risen to about a third of the population.[6]

One of the reasons, surely, is that there are so many aspects of the climate that we don't entirely understand. Take aerosol, for instance, by which I do not mean your underarm deodorant, but small particles put into the atmosphere by fires, volcanic eruptions, sea spray, dust storms, industrial processes and jet aircraft emissions. It's long been known that these affect the planet's temperature. The problem is we're not sure how much, yet they could be deeply significant.

Aerosol particles provide something for water droplets to condense around and thereby form clouds (indeed, if there weren't any aerosol particles in the air there wouldn't *be* any clouds). The whiter the cloud, the more sunlight it helps reflect back into space, cooling the planet. (Everyone who's flown over banks of clouds will know how startlingly bright they can be.*) In a 2006 paper published in *Science*, François-Marie Bréon at the Laboratoire des Sciences du Climat et de l'Environnement suggested human aerosol emissions "may increase cloud cover by up to 5%,

---

*Nothing is simple in climate science. Some clouds actually heat the planet, notably high thin ones. It's the lower thick ones that reflect solar radiation. Inevitably, there is debate among climatologists about how the ratio of "warming" to "cooling" clouds will change with global warming.

resulting in a substantial net cooling of Earth's atmosphere."[7] Indeed, one technique proposed to offset global warming is a fleet of "cloud seeding ships" that will scoop up seawater and force it through a system a bit like an inkjet printer to place tiny droplets of just the right size into the air, around which clouds can form. The argument is that more clouds could temporarily offset the heating effects of global warming, giving us longer to decarbonize our economies.[8]

We also know that some aerosols can reflect sunlight away from the planet by themselves. In 1991, the eruption of Mount Pinatubo in the Philippines sent roughly twenty million tons of volcanic ash twelve miles high into the atmosphere, and average global temperatures went down by about half a degree centigrade the following year.[9] The ice across Hudson Bay melted almost a month later than normal, and polar bears, who feed and give birth on the ice, had a greater number of healthy cubs that summer (offspring dubbed "Pinatubo cubs").[10] Then again, other aerosols like soot *absorb* sunlight and have a warming effect.[11] Some believe that this kind of "black carbon" "may be the world's second-leading cause of global warming after CO_2."[12] As Mike Berners-Lee says in his book *How Bad Are Bananas?: The Carbon Footprint of Everything,* "The good news about black carbon is that it lasts only a few days in the atmosphere. In other words, if we reduce the amount we create, the benefit will be instant. Hence some experts think that reducing black carbon pollution should be a number one priority in tackling global warming."[13] The problem is we still don't know how much aerosols affect temperature, mostly because we don't know how much of each sort is in the atmosphere. In January 2010, a *Nature* article entitled "The Real Holes in Climate Science" summed up the situation by saying estimates of the net aerosol effect vary by an order of magnitude.[14] As I said, there's still a lot of debate to be done.

Another gap in the climate change picture is understanding where the carbon we put into the atmosphere actually goes. We know how much we emit, and we know how much stays in the atmosphere (about half) but we're not sure exactly where the *other half* is going. It's certainly being absorbed by the land and seas but in what quantities and where is still not fully resolved.[15] It was hoped that NASA's Orbiting Carbon Observatory would answer this question, except it crashed into the waters around Antarctica in February 2009.[16] A replacement is on order, but with a price tag of $278 million it's not something that can be turned around in a weekend.*

James Lovelock says, "The great climate science centers around the world are more than well aware how weak their science is. If you talk to them privately they're scared stiff of the fact that they don't really know what the clouds and the aerosols are doing. They could be absolutely running the show. We haven't got the physics worked out yet."[17] In climate science circles, this is an extreme view and many would challenge Lovelock, although as a grandee of climate science he's always worth paying attention to. His comment does, however, demonstrate the point (probably in a far too blunt, inflammatory and off-the-cuff way for many climatologists) that we don't have a fully coherent view of what's going on. Or to put it another way, trying to build a comprehensive understanding of the climate is a bit like trying to follow my mother in conversation—you have to recognize it's an entirely nonlinear system.

When former U.S. secretary of defense Donald Rumsfeld gave his now famous assessment on the war in Iraq, he could have easily been talking about climate science:

> As we know, there are "known knowns": there are things we know that we know. There are "known unknowns": that is to say there are things that we now know we don't know. But there are also "unknown unknowns": there are things we do not know we don't know.

Stronger as an exercise in logic than public speaking, it's not a bad way of summarizing the climate science debate.

---

*Not all is lost. The Japanese GOSAT designed to monitor greenhouse gas emissions during the first Kyoto protocol enforcement period from 2008–2012 is providing some data.

The biggest "known known" is the amount of carbon dioxide we're putting into the atmosphere along with the science that proves that CO_2 has a warming effect.* "Known unknowns" include the impact of aerosols on temperature and exactly where the carbon is going. "Unknown unknowns" are things like climate "tipping points" we're not even aware of that might suddenly trigger great leaps (or falls) in temperature.

The problem for many skeptics is, if someone's predicting a possible apocalypse you want them to be one hundred percent sure about it. We're told our cities may flood, whole nations may become uninhabitable, that the carrying capacity of the planet will plummet and wars will erupt as we fight for ever scarcer resources. Then we realize that climate science isn't exact. To the nonscience literate this makes the whole enterprise sound rather unsure of itself—and given the gravity of the possible outcomes, many people rather expect a stronger-worded case. (Ironically for many scientists, their consensus on the climate change threat represents the *strongest* wording they've ever collectively come up with.)

Then there's the fact that the most important figure used in the climate change argument seems intuitively nonthreatening—sure CO_2 levels have gone up, but by a hundred parts *per million*. And the rise since the Industrial Revolution? About a thousandth of a percent of the entire atmosphere. As Australian MP Bob Katter said to his Parliament, "Are you telling me seriously that the world is going to warm because there are 400 parts *per million* of CO_2 up there? If you know anything about science, you realize how utterly preposterous that proposition is, how absolutely ludicrous and ridiculous it is."[18] Bob, unlike a scientist, or indeed a whisky drinker, finds it hard to understand that a small amount of something can have a large effect. On top of all this, we're used to hearing about the next

---

*It's actually pretty easy to demonstrate the warming effect of CO_2 with a few test tubes, some stoppers, a couple of thermometers and ingredients you can find in your kitchen. You can make your own CO_2 by adding some acid (say, lemon juice or vinegar) to baking soda. (Note: You do of course make carbon dioxide yourself every second—it's in every breath you exhale—but trying to trap this in a test tube just makes you look stupid.) Once you've made your CO_2, add it to some air in a test tube and quickly seal. Now seal another test tube without added CO_2 and stick both out in the sunshine for twenty minutes. Measure the temperature of each. *Voilà!* One is hotter than the other.

apocalypse almost as often as Amy Winehouse goes drinking. In the last fifty years we've been warned about numerous soon-coming catastrophes including a new ice age, nuclear Armageddon, the millennium bug, SARS and various flavors of a worldwide flu pandemic, none of which have so far happened. We've become weary and a little cynical of what feels like yet another "end of the world as we know it" scenario.

Finally, there's winter. As Jon Stewart pointed out to climate change campaigner and former U.S. vice president Al Gore on *The Daily Show*, it still gets cold every winter and people wonder, "Should I believe this Al Gore, or my rock-hard nipples?"[19]

Combined, it's not hard to see why so many people look at global warming as an unfathomable, shifting sideshow to their daily lives. "Come back when you've got a coherent story," they think, and who can blame them? It's not so much that there isn't a strong and urgent case for the dangers of man-made global warming, just that for many people it hasn't been made yet. Others might argue that because global warming is a problem that builds slowly over a long time and that dealing with it requires some fundamental changes to the way we organize our societies, it's just too difficult for people to get their heads around. Or perhaps you think that grasping the problem requires a level of scientific literacy that some of us just don't have, or aren't inclined to develop—so people simply "switch off." However you want to slice it, the reality is that for many people there is a very strong emotional reason to adopt a "wait and see" attitude.

About once a month I like to play poker with a few of my neighbors. It's an excuse to have a few beers, exchange stories and lose most of our money to Brian. We tend to play either Texas Hold 'Em or Omaha— games where an extra card is revealed on each round and players bet on how their hand is shaping up. One of the reasons Brian tends to take home more than me is that I'm a bit of a "wait and see" player. I'm rather optimistic about the cards still in the deck. If I have a two and a five of hearts I'll hang on to the end, waiting for three more hearts to give me a flush. More often than not I should fold. The trouble with this kind of optimism is I end up betting more money—and generally losing it. Brian is the opposite. If his hand looks weak, he folds early. He bets high when the odds are good, bails when they're bad—and occasionally, because we know this, he bluffs the hell out of us and wins with a pair of threes. (Bastard.)

My worry is that a "wait and see" attitude in relation to the climate could be the worst mistake we ever make. Just because the climate models are incomplete doesn't mean that all those worried scientists are wrong. There are some pretty convincing "known knowns." The temperature *is* rising. Arctic ice *is* receding. CO_2 concentrations keep going up. What if the Intergovernmental Panel on Climate Change's scientifically worded assessment that it is "very likely" man is warming the planet becomes, "in retrospect it was a certainty"? How much do we want to bet on this? Our major coastal cities? New York? Mumbai? Singapore? London? Tokyo? Or perhaps a few low-lying countries like the Maldives and Bangladesh? What about those billions of lives James Lovelock believes will be forfeited? Do we bet high or low? What are the stakes? Are we confident enough that those scientists are wrong? Are we willing to wait for the flush? I guess, at the end of the day, I'm not. Despite the gaps, the basic science is convincing. I may be happy to bet and lose twenty quid to Brian once a month in return for good banter and some quality Scotch, but I'm not sure I'm happy to bet and then lose billions of my fellow humans to a world with a much lower carrying capacity. "Climate change doesn't kill you," says Stewart Brand. "Other people fighting over the reduced resources caused by climate change kills you."[20]

From the air, you can see just how vulnerable New York is to sea level rise. As my flight approaches LaGuardia airport I find myself thinking that, to be convincing, skeptics actually need to show a *much stronger* hand than the climate change lobby. They need to *conclusively prove* global warming *isn't* man-made. Over eight million people in NYC alone need to know what to do. Leaving them to drown on a hunch isn't an option.

Opinion isn't going to cut it. It requires solid reasoning backed up by research. It needs to convince ninety-six percent of active climate scientists.[21] It is not enough to *cast doubt* on man-made global warming. Skeptics need to conclusively nail that sucker. Because when it comes to the planet (as opposed to poker), I think I'd prefer to play safe. The worst that can happen if we're wrong about global warming is we lose some economic growth and accelerate the development of clean technologies to replace fossil fuels when they run out. The worst that can happen if the skeptics are wrong is that most of us die. In the words of *Dirty Harry*, if you're a skeptic "you've got to ask yourself one question: 'Do I feel lucky?'" Well, do you?

But even if you are a skeptic, stay with me. The next stages of my journey will show me things that are good for the planet and our economies whether you accept man-made global warming or not.

I land as dusk is descending and take a taxi to the up-and-coming but knowingly shabby neighborhood of Long Island City. I'm staying with arguably my most egg-headed of friends, Colin, a neuroscientist whom I met via a flat-sharing website when I needed someone to share my rent in London. Colin is trying to get to the bottom of how memory functions in order to throw some light on diseases like Alzheimer's. This makes him a bit of a hero in my world.

Colin is actually in San Diego tonight being interviewed for a new job so I have his place to myself. Scattered about the apartment are research papers with titles like "Hippocampal CA3 Output Is Crucial for Ripple-associated Reactivation and Consolidation of Memory." What's different about seeing this sort of thing today as opposed to when we lived together is that now I want to pick them up and understand them. (Previously, our scientific discourse stretched as far as designing the perfect breakfast sandwich.) One trip to Harvard, MIT and Eric Drexler's favorite coffee shop and I may be becoming a geek.

The next day is all sunshine and blue skies. I take a trip to Manhattan and sit overlooking the Hudson River in Rockefeller Park, preparing for my next meeting with some super-brains. I'm going to meet geoscientist Wallace Broecker and Klaus Lackner, a German physicist who may be to excess CO_2 what Dirty Harry was to criminals. I'm momentarily distracted by a boat emblazoned with the words "America's *only* gay sailing tea dance." Well, I guess it's a niche market.

Colin returns from San Diego in a dilemma. The job looks perfect, but he's not sure about leaving his beloved New York. "I understand," I say. "There's no gay sailing tea dance in San Diego."

"Right," he says, and falls silent.

In climate science circles Wally Broecker is a combination of rock star and institution, the geoscientist equivalent of Willie Nelson. As I set off the next morning to meet him and Klaus, I'm feeling a mixture of apprehension and excitement. Broecker has been described as "the grandfather of climate science" and "one of the world's greatest living geoscientists."[22] He's the recipient of a brace of awards, which if listed here would make your eyes glaze over, but include the U.S. National Medal of Science and the Tyler Prize (awarded annually "for environmental science, energy and medicine conferring great benefit on mankind"). "I think the greatest pleasure is beating nature to one of her secrets," he tells me when we meet. "I'm an 'inverse engineer' in a sense. We have an Earth system and I'm trying to work out how it's put together." On his cabinet of trophies he remarks, "Longevity helps."

Among scientists he's probably known best for his work on "thermohaline circulation" (the ocean's "conveyor belt") but more recently has become known as the first person to use the words "global warming" in print. "If my career has boiled down to *that* it's a big failure," he says.*

---

*Wally remarks, "That was in an article I wrote for *Science* in 1975. It was in the title so I guess any computer search can pick it up, I don't know if it's really true that I was the first one, but I certainly didn't use it with the intention of coining something—it just happened to be the likely thing to say."

My first stop is Klaus Lackner's tenth floor office at Columbia University's Manhattan campus (the building, inexplicably, smells of ham sandwiches). In person Klaus is tall, with smooth features that if it weren't for his gray hair could fool you into thinking he was in his forties rather than his fifties. His facial expression seems in constant indecision about whether it is going to help him articulate a deep thought or a joke, which combine to give him an approachable air. As it turns out, when Klaus talks scientifically he has an easy delivery and a smile on his face, and when he tells a joke he has a Germanic precision and looks slightly concerned.

Klaus is giving me a lift to meet Wally at the leafy campus of the Lamont-Doherty Earth Observatory, a thirteen-mile journey north. The observatory is dedicated to studying the planet at a "big picture" level, understanding Earthwide systems and how they interact. In the short time I have in his beat-up car it becomes abundantly clear he has a deeply analytical mind, inherited perhaps from his lawyer father who worked to build a fair judiciary in Germany after World War II. But as well as a scientist's need for clarity, there's an empathy for the ambiguities of the human condition. As we cross the George Washington Bridge, Klaus recalls in his perfect but slightly accented English the traumatic effects of conflict on his older relatives. Perhaps this gives some clue as to why Klaus isn't just an "ivory tower" theoretician. Like his father, Klaus is determined to *do*, not just think. "I have an engineering bent," he says. "So I'm not just looking at how and why things work, but how one can *make* things work. I'm very much interested in how to build things." The world may one day be very glad that Klaus is like that.

I use the journey to ask how is it that a gas that exists in parts *per million* in the atmosphere can have such a dramatic effect on global temperatures.

"Well, in relative terms it's not small," says Klaus. "We've gone from two hundred and eighty parts per million to three hundred and eighty, so the increase is more than a third. If you drink three beers instead of two, that makes a difference. But the main reason it matters in the climate is because it's such a *potent* greenhouse gas." In other words, a tiny amount can make a huge difference—a bit like putting a drop of Tabasco in your brother's contact lens solution.*

"What makes CO_2 so potent?" I ask.

"It's a big player when it comes to infrared absorption in the atmosphere," explains Klaus. Well, yes indeed. But what does *that* mean? It turns out it's all to do with wavelengths of energy.

Energy is a hard concept to visualize. We can see or feel its *effects*, but we can't see *it*. You just can't visualize "energy" in the same way you can conjure up the image of water streaming from a tap. Given that energy itself is a pretty esoteric concept, talk of wavelengths of energy can be even trickier to grasp.

Most of us can get our heads around wavelengths of *sound*. Nearly every hi-fi and radio has "tone" or "bass" and "treble" dials to boost higher or lower frequencies. If you've got a really fussy hi-fi, you can even fiddle with a "graphic equalizer" to give precise volume boosts to the specific wavelengths you want more of (my brother, for instance, likes enough bass to challenge the integrity of his colon). We understand that low noises, like the bass riff from "Superstition," have longer wavelengths than the masterful sax part, which we hear as a higher set of sounds (with shorter wavelengths).

It's easy to understand that some materials absorb different wavelengths of sound better than others. The walls between me and my neighbors, by example, are pretty effective at taking out shorter sound wavelengths (high sounds), but not so good at filtering out the longer ones. As a result I'm intimate with the bass lines of their entire record collection. Klaus explains that CO_2 absorbs certain wavelengths of *heat energy* coming up from the planet's surface, just like my wall likes to absorb

---

*I find another example of a small amount of material having a big effect when Colin introduces me to his boss, neuroscientist René Hen. "What got you into neuroscience?" I ask, and he laughs. "Um . . . it was my experience with magic mushrooms a long time ago. The idea a tiny amount of this discrete compound could have such a powerful behavioral effect was interesting. You take half of a mushroom and you get effects that are pretty profound and last for hours . . ." I bet he didn't say *that* in his interview.

certain wavelengths of sound energy coming from next door. However, once CO_2 has absorbed energy, it likes to *re-emit* it, and a good proportion of that heat goes *back down* (to the Earth's surface) rather than *up* (escaping into space). The result? More CO_2 makes a warmer planet.

"CO_2 absorbs wavelengths of energy that other greenhouse gases don't," says Klaus. To paraphrase an old beer advert: CO_2 reaches the energy wavelengths other greenhouse gases can't. In doing so it plugs some "gaps" that usually let heat energy out into space. One of those other greenhouse gases is water vapor.

An atmosphere rich in water vapor is a large part of the reason we enjoy a habitably warm planet. The moon provides an example of what can happen when there isn't a lot of water vapor or other gases around to help form an atmosphere. In fact, the only place in the solar system with less atmosphere is Vadim's nightclub in Swindon.

Water vapor is the largest absorber of heat energy coming up from the Earth's surface and is without doubt the most significant greenhouse gas. Its concentration in the atmosphere, however, is regulated by the planet's natural processes—and there's nothing we can do about it. If you pumped a whole load of extra water vapor into the atmosphere it would soon condense into rain and snow. Conversely, if you sucked some of it out, evaporation from the oceans would quickly replace it. The U.S. National Center for Atmospheric Research estimates that every year 473 *trillion* tons of water evaporate from the ocean, and a further seventy-three trillion evaporate from the land. (This compares to about thirty billion tons of man-made CO_2 emissions.) Going after water vapor as a way to combat global warming would be like trying to run after a sumo wrestler covered in baby oil. Despite the bulk of the target, it'd ultimately slip through your fingers.

"The water vapor is uncontrollable," explains Klaus as we drive north. "It comes and it goes. But the CO_2? We put it in, and it stays."

David Archer, a University of Chicago oceanographer and author of *The Long Thaw: How Humans Are Changing the Next 100,000 Years of Earth's Climate*, writes, "The lifetime of fossil fuel CO_2 in the atmosphere is a few centuries, plus twenty-five percent that lasts essentially forever. The next time you fill your tank, reflect upon this."[23]

Eric Drexler told me, "People think of carbon dioxide like a pollutant. They think that when you stop producing it, it goes away, like smoke in the sky. But with CO_2 the time taken for it to return to the baseline level is measured in centuries. If you were to shut off CO_2 production today, we'd continue to warm."

Importantly, there's a direct link between the temperature of the planet and the concentration of water vapor in the atmosphere. The hotter we get, the more water vapor is taken up by the air. This means that adding CO_2 to the atmosphere doesn't just warm the planet by trapping heat energy *in and of itself*, but encourages *more water vapor* into the air too. This means water vapor acts as a big *amplifier* of any warming effect CO_2 has—your baby-oiled sumo wrestler just got bigger. This is a key reason why CO_2 is so potent. When it gets into the atmosphere, it invites a whole load of warming water vapor to come with it (although it's *another* ongoing debate in climate science as to how much water vapor is taken up by the air as the temperature changes).[24]

I'm seriously considering installing some soundproofing in my flat. I'm looking for something which will absorb those longer sound wavelengths that come through the walls and therefore allow me to study my "How to Kill Your Neighbors and Get Away with It" correspondence course in peace. Just as I can buy soundproofing for my flat to absorb specific wavelengths of sound, so the planet can get extra "heat energy proofing" for the atmosphere in the form of additional greenhouse gases. If an evil genius could nip down to some planetary hardware store and say, "What have you got that's easy to install, durable and will warm up the planet by a few degrees by Friday?" they would likely receive the reply, "CO_2 mate. It's cheap, lasts for ages and better still, it comes with a bunch of free water vapor too. They'll be roasting in no time."

Klaus and I have arrived at the Gary C. Comer Geochemistry Building, where Wally is waiting for us.

If Wally were a Muppet, he'd be Waldorf, one of those two old guys in the balcony—smart, funny and ready to point out, without remorse, the shortcomings of things placed in front of him. He has a charmingly curmudgeonly manner with little time for niceties with visiting authors ("What are you doing here? You're writing a book? Oh yes, I have some vague recollection"), and at seventy-eight retains a good sense of mischief. He recalls playing a prank on fellow Lamont scientist George Kukla, jacking up his car and placing it carefully on cinder blocks just a mite bigger than the normal gap between the chassis and the ground. When George tried to drive off, his wheels spun impotently. The normally calm research scientist lost his cool, not least because he was entertaining "a very important man from China, one of the first Chinese visitors that came here." Wally laughs. "We asked him whether the Chinese did pranks like that and he said: 'Only small children.'" He smiles and laughs again.

When Wally talks about climate, people listen. He insists the warming we're seeing now is fundamentally different to historical shifts in the climate. "It's bigger and faster," he tells me. Which naturally prompts the question "What are we going to do about it?"

The options generally presented are to do nothing (this from the global warming skeptic camp), cut CO_2 emissions (this from optimistic climate activists) or engineer countermeasures to produce some kind of counteracting cooling effect. This last option is called "geo-engineering," of which the wackiest idea is launching large mirrors into orbit. Geo-engineering is usually suggested by *pessimistic* climate activists. But there is another option. Invest in Klaus Lackner.

Klaus and his colleagues have built a machine that strips CO_2 out of the air. Or to put it another way, on one side of Klaus's machine is air that contains current levels of CO_2 and on the other is air with roughly the same amount of CO_2 as was present before the Industrial Revolution. "We spent the last five years in Tucson, Arizona, proving that this works," says Lackner. The "we" in question is Klaus and two brothers brought to the table by Wally: Allen and Burt Wright, who are unrelated to the famous flying Wright brothers, but who might also go down in history for their experiments with machines that interact with the air.

If the promise Klaus sees in his technology comes to pass, it can begin to *reclaim* the CO_2 we've been emitting. It's not about reducing what we emit, it's about extracting what we've put there, treating emissions in the same way we treat sewage. This could be a crucial component in a CO_2 processing infrastructure for the planet.

It isn't the whole answer (although with enough of Lackner's machines it arguably could be). "If you've built a coal plant that captures CO_2 right out the smokestack, I can't compete with that," says Klaus. But even if every power station suddenly became a zero-emitter of carbon tomorrow, there are plenty of other places pumping it out, especially in the transport sector, which accounts for nearly a quarter of the world's emissions. "An airplane has a hard job running on electricity," he says.

The Lackner/Broecker position is that creating waste isn't necessarily a bad thing. *Not dealing with it* is the bad thing. Nobody suggests you stop going to the toilet, but we did install sewage systems. In the UK it was actually "The Great Stink" of 1858 (during which the smell of untreated sewage assailed central London, and crucially Parliament) that finally convinced legislators to invest in sewers. You won't find many people arguing against sewers today (and if you do, don't accept a dinner invitation from them). Lackner's carbon scrubbers are one option for treating our "carbon sewage." It makes impossibly simple sense. What's more, Klaus's machines aren't just an idea on paper. Lackner's self-confessed "engineering bent" means that, along with the Wright brothers, he's already built some laboratory-scale working prototypes.

Carbon scrubbers aren't a new invention. They've been used for decades, for instance, in submarines to keep the air breathable. Until recently, however, the prevailing wisdom was that such scrubbing technology could not be adapted to remove the relatively small proportion of CO_2 in the atmosphere without using huge amounts of energy. Indeed, a 2005 special report from the IPCC dismissed Klaus's work with a single line: "The possibility of CO_2 capture from ambient air (Lackner, 2003) is not discussed in this chapter because CO_2 concentration in ambient air is around 380 parts per million, a factor of hundred or more lower than in flue gas."[25] Or in other words, you'd be nuts to try and find the CO_2 needle in the atmosphere haystack. Anything that might work would take up too much energy (and thus add more CO_2 to the atmosphere than it removed). Wally initially had reservations too. The first time he saw Lackner talk, he thought the German was nuts ("energetically nuts," to quote him directly). "Then we had more time to talk, and I immediately tried to hire him." Lackner set out to prove his methods could remove CO_2 using low energy levels, and Wally was right behind him. That's because Wally believes that a plan that only focuses on cutting emissions "is going to kill us."

"People say we've got to stop using fossil fuels," he says. "We look at it the other way. They're going to be burned. The world's going to need energy, the developing nations are going to use more energy, so we better damn well figure out what to do about it. We can't change over fast enough," says Wally. "The world leaders still don't really get it. That's why we need air capture."

So how do Klaus's machines work?

If you were a chemist you'd know that sodium carbonate will, given almost no provocation, react with CO_2 to create sodium *bi*carbonate (or baking soda to you and me). The key component in Klaus's machines is a hanging gallery of strands of a "sorbent" resin, impregnated sodium carbonate, which react with the CO_2 in the air flowing over them, the captured CO_2 helping to create baking soda.

Capturing CO_2, though, is only one half of the job. Somehow you've got to get the CO_2 *off* the sorbent if you want the apparatus to be reusable and therefore cost-effective. Restocking the whole shebang with a new supply of sorbent resin makes things prohibitively expensive and energy hungry. This is where Lackner's resin comes into its own, by doing something that even Klaus admits is counterintuitive. In the presence of water the resin changes its affinity for CO_2, shedding its recently collected bounty. The "collection" reaction takes a reverse step. Sodium *bi*carbonate becomes sodium carbonate. What this means is that if Klaus pumps water vapor into his machines, CO_2 from the sorbent will "fall off" the resin, allowing the whole apparatus to be reused. Condensing that vapor allows the captured CO_2 to bubble out the top, in the same way CO_2 bubbles rise to the top of champagne.

There's a kind of sweet poetry to one greenhouse gas (water vapor) collecting another (CO_2). After all, one of the problems with CO_2 in the atmosphere is that it encourages more water vapor into the air, thereby amplifying the warming effect. Here, thanks to the chemistry of Lackner's sorbent, the opposite is happening. Water vapor is being used as part of a process to take CO_2 *out* of the air.

Of course, once you've captured the CO_2 you've got to store it somewhere (a process called "sequestration") and make sure it doesn't find its way back into the atmosphere. One option is pumping the gas into rocks: "Ultramafic" rocks, which have high concentrations of magnesium that react with CO_2 to create magnesium carbonate. Klaus tells me Oman alone has enough of these rocks such that there isn't enough coal left to burn that would overwhelm them as a CO_2 store. Another option is pumping CO_2 into basalt rocks to make limestone—an alternative being investigated by another Columbia scientist, David Goldberg.[26]

"Can the chemistry of the sorbent be improved further?" I ask.

Wally jumps in with a guffaw. "They don't know how the chemistry works!" he exclaims with boyish joy.

Klaus isn't sure why water vapor makes his resin give back some of its CO_2. "I can tell you for sure *what* it does. That we can see. But at the moment I can only speculate *why* it does it. I've a good theory, and we will prove whether I'm right or wrong. One of the reasons I'm excited about where we are right now is we are setting up experiments to understand the chemistry. Once we've done that, we can *engineer* the chemistry. So I guarantee you these machines will get better."

In his office Wally points to a sealed tube (next to a can labeled "Dr. Fozze's Fart Beans") that contains some of the first CO_2 captured by one of Lackner's early prototypes. Klaus now says that for every twenty CO_2 molecules his machines will put in the atmosphere (if they're powered by electricity generated from fossil fuels), they'll take out a hundred. And he's just at the beginning of his journey. With investment, experience and improved manufacturing, that ratio will get better. (Klaus already has a long list of improvements he wants to research.) So it seems crazy that Klaus has struggled to raise the twenty million dollars he estimates he needs to turn his prototypes into a blueprint for a mass-manufactured unit.

"The fact that Klaus has trouble raising money is absurd," says Wally, bristling. In fact, when I meet them, the two men are reeling from a recent decision by the federal government not to fund a research hub dedicated to carbon capture and storage.[27] Klaus's requirements amount to just 0.000025 percent of the $787 billion the U.S. government pulled out of the hat for its American Recovery and Reinvestment Act of 2009— money for stimulating the economy out of the economic crisis (or one quarter of one ten-thousandth of a percent).

It's ironic that when it came to saving the financial system, governments around the world couldn't move fast enough. "It's the platform our economies run on," they cried in order to justify swift, decisive and taxpayer-crucifying action. Yet there is another platform all the banks run on. It's called the atmosphere, and the social and financial implications of global warming will do more to hamper Wall Street than anything they've done to themselves. When, I wonder, did a human-friendly atmosphere not become an infrastructure investment? A back of a napkin calculation suggests we could build enough of Klaus's machines to reclaim all the carbon we pump into the atmosphere each year (and start to reclaim the backlog) for the equivalent of a three percent tax on car prices for the next ten years.

Why does Klaus think it's been hard to find the necessary funds?

"I think our nature is that if the crisis is tomorrow we'll jump, we'll have the adrenaline to do whatever it takes to solve the problem," he says. "If I told you fifty years ago that what was happening in banking would lead to a meltdown . . . would we have done anything? We are not good at thinking beyond a fifty-year time frame." Maybe we need another "Big Stink"?

In the meantime Klaus can explore another avenue. CO_2 has lots of industrial uses. For instance, it's used to improve yields in commercial greenhouses. All plants require atmospheric CO_2 to live; it's the source of the carbon they convert into sugar and other carbohydrates through photosynthesis (and yes, this is another one of those feedback loops in the climate we don't fully understand—the extent to which adding more CO_2 will produce more plant growth on the planet).

CO_2 makes fizzy drinks fizzy, powers pressure tools, is a propellant for aerosol cans and an ingredient in fire extinguishers. It can be used as a tool to create "pure" atmospheres for welding, an aid for the construction of nanoparticles used in medical drugs, a refrigerant, a raw material to make polymers and plastics,[28] and even as a theatrical effect (dry ice).

"The U.S. consumes roughly eight million tons of commercial CO_2 a year," says Klaus. "So you could push this forward without having government support. You start small, selling CO_2 into the market, improving your technology, and you can be ready before the coal plants have figured out the best way to capture CO_2 at source."

There's no way around it, Klaus is good news for the planet. Even better news is that he isn't the only one developing machines that take carbon out of the air. "I convinced David Keith [a renowned climate scientist] that this air capture stuff works so he now has a competing effort," says Klaus. Peter Eisenberger, also at Columbia University, is attacking the problem as well. Another promising idea is called "Solar

Thermal Electrochemical Photo Carbon Capture," which has been demonstrated by Dr. Stuart Licht and colleagues at George Washington University. The team claims that if the technology was scaled up to cover seven hundred square kilometers, the process could "remove and convert all excess atmospheric CO_2 to carbon" in just ten years.[29] (That's a carbon capture plant roughly the size of Micronesia). The more people working on technologies to take back the CO_2 we're putting into the atmosphere the better. David Keith told the *Calgary Herald*, "who will actually take it forward is now a horse race."[30]

"Let's imagine a world in which we suddenly have lots of Lackner scrubbers and you bring the levels of CO_2 in the atmosphere down to preindustrial levels," I say. "Does the planet start cooling almost immediately? Does the warming stop?"

"We've warmed up the ocean and that will act as a damper on cooling," says Wally. "The ocean is holding back the warming of the planet too, because it sucks up a lot of the heat from the atmosphere. But as we cool the planet, the ocean's going to give that heat back, and slow down the cooling process."

"The land would give its extra heat back in a couple of days," explains Klaus, "but the oceans will take decades, although you will see it going back down quite fast in the beginning."*

I ask Klaus how he goes about convincing people he's onto something.

"The problem I've found (and it's getting bigger all the time) is that I'm suspect to both sides of the debate. The people who make energy or are into coal feel I'm trying to stop them—because I'm saying you've got to take climate change seriously and business as usual is not acceptable. On the other side you have people who have some idea of what 'being green' means—and allowing people to use fossil fuels is not acceptable to them."

---

*Of course, we can't suddenly snap our fingers and fill the world with enough of Klaus's machines to start instantly offsetting all our carbon emissions. "You can't do it overnight, but I do believe you could do it in a decade once you know what you are doing," says Klaus. "So, you'd have a thirty- to forty-year delay until you are back to normal."

"Do you think traditional environmentalists are part of the problem now?" I ask.

Wally snorts. "Oh yeah!" Wally comes into conflict with "traditional greens" is their objections to experiments relating to CO_2 sequestration. "I get very upset. They strongly object to even *doing experiments*," he says.

Is he optimistic we can solve the CO_2 problem?

"It'll be solved. The question is where will CO_2 levels get to before it's solved?"

Klaus agrees: "I'm optimistic that ultimately it will be solved. But my view of human nature is that we will not solve it until we get seriously goosed."

"Maybe in twenty years, when the impacts become obvious, we'll get serious," says Wally.

"But let me give you an optimistic view," says Klaus. "Back in the nineties I was asked, "How do you see this moving forward?' and I said, 'In this decade, the nineties, you will see scientists thinking about it and not much more. The next decade there will be a big political debate and not much more. The decade where steel starts to go into the ground is 2010 onwards. And people get really serious about it between 2020 and 2030.' In a way, we are on that track."

Wally announces that he has to go for a beer with George Kukla (the car prank obviously long forgotten), signaling the end of our conversation. Before he goes, he points to a picture on the office wall which shows him in a group of honorary degree recipients from Cambridge University. Standing near Wally is Microsoft founder Bill Gates.

"Why isn't he giving you some money?" I ask.

"I did send him some stuff but didn't get a reply," says Wally, and heads off for his drink.

As Klaus and I make our way back to his car, we pass a huge furry pink and blue toy snake pinned to the wall outside Wally's office. Underneath it a piece of paper bears the words, "I am the climate beast and I am angry!" It's arguably Wally's favorite metaphor. His assertion that "The climate is an angry beast and we are poking it with sticks," is one of the most quoted summaries of our problems with carbon dioxide (CO_2 being the "stick" in question).

Klaus is giving me a lift to the Dobbs Ferry train station for my trip back to Manhattan. As I get to his car I turn to him and say, "You must be excited?"

"Oh yeah," he says. "Oh yeah."

As my train makes its early evening journey along the east bank of the Hudson, something is scratching at the back of my mind. I know I've heard something on my trip about CO_2 that's significant to the conversation I've just had with Wally and Klaus but it's eluding me. I pass through Hastings, Greystone, Glenwood, and then, as we pull out of Yonkers, I remember: Joule Biotechnologies, a company George Church had mentioned to me in passing (and one of the many organizations to which he is a scientific adviser). The company uses genome engineering to make strains of bacteria that, under sunlight, consume CO_2 and excrete diesel and ethanol fuels. I whip out my iPhone and find their website, silently thanking Vint Cerf that I can.

"Joule efficiently captures sunlight to produce energy in liquid form," it says, claiming its "use of waste CO_2 as a sole feedstock creates the potential to deliver virtually unlimited quantities of fuel." In April 2010, MIT's *Technology Review* named Joule one of the ten most innovative companies in the world, noting that it bypasses the major criticism leveled at biofuel production—the amount of crops and land needed to grow biofuel feedstocks such as corn, sugar beets or switchgrass. As the U.S. Energy Information Administration (which produces independent statistics for the U.S. government) remarks, "The greatest challenge facing biofuels production is to secure sufficient raw material feedstock for conversion into finished fuels. Production of biofuels requires significant land use dedicated to the growth of feedstock crops, and land prices could represent a significant constraint."[31]

Sugar beets and corn produce about five hundred and eighty gallons of ethanol per acre according to the Renewable Fuels Association, a lobbying group for the U.S. ethanol industry.[32] Estimates for algae biofuels vary wildly but even those on the conservative side suggest it can do about ten times better, with per acre yields of five thousand gallons of oil (which then needs to be refined).[33] Florida-based Algenol Biofuels has developed a "hybrid algae" that, via photosynthesis, converts CO_2 *directly* to ethanol. They claim a yield of six thousand gallons per acre[34] and received one of fifteen $25 million grants issued by the U.S. Department of Energy to help develop a pilot "biorefinery" in Texas.[35] Joule claims their system is *already* delivering six thousand gallons of ethanol per acre, and this figure will rise to twenty-five thousand per acre once they start full-scale commercial production.[36]

Ethanol has struggled to get off the ground as a fuel largely due to costs of production (and linked concerns over land use), but with innovations like Joule's and Algenol's, that could be about to change. The U.S. consumed 137,800,000,000 gallons of gasoline in 2008. If all of that had been sold via gas stations (there are 162,000 in the U.S.) it would mean, on average, each outlet would have distributed just over 850,000 gallons that year.[37] Assuming all gas stations are born equal (of course they're not) each one would need to find between thirty and forty acres of land to make enough fuel using a system like Joule's (this takes into account the extra ethanol needed to compensate for the fact it delivers slightly fewer miles per gallon than gasoline). That's just one-sixteenth of a square mile. Couple Lackner's CO_2 capture technology with a Joule or Algenol system, and you could build an ethanol or diesel factory nearly anywhere where there's sunshine. We could literally pull our fuels out of the sky. Of course, the burning of those fuels would put the CO_2 back into the atmosphere but, crucially, such a system wouldn't *add* any carbon dioxide to the sky, as our current use of fossil fuels does. It'd be what eco-geeks call "carbon neutral."

Even if you don't buy the need for carbon neutral fuels, I'm guessing you're in favor of a more peaceful world. Today nearly ninety percent of oil reserves are held by just thirteen countries, which makes the rest of the world rather dependent on them.[38] The United States, for instance, imported fifty-seven percent of its oil in 2008.[39] As reserves dwindle, the opportunity for conflict is obvious.* Beyond the possibility of making any country that successfully adopts these technologies energy-independent, they also eradicate the sort of man-made disaster that's devastating the U.S. coasts of the Gulf of Mexico as I write, the result of the Deepwater Horizon oil rig explosion.

On top of all that, the energy infrastructure could stay largely the same. Gas stations would look like gas stations today; so would cars. There's no need to radically reengineer our transport infrastructure as would be necessary in shifting to hydrogen-powered or electric vehicles. Already there are eight million "flexible fuel vehicles" on American roads that can run on a gasoline/ethanol blend (many are "E85" vehicles that can run on blends of up to eighty-five percent ethanol and fifteen percent gasoline).[40] General Motors has cautiously committed "to making 50 percent of production flex-fuel capable" by 2012.[41] Ford, Chrysler and Toyota all offer E85 cars. In fact, Ford is returning to its roots—its famous Model T, which went into production in 1908, could run on ethanol, gasoline, or a blend of both.

So why not fuel made from CO_2 taken out of the sky? Already we have working technologies that can take CO_2 from the air and organisms that can turn CO_2 into liquid fuels. Klaus may not prevail; Joule might fail; Algenol might be a flash in a pan. I hope not. But even if they are, somebody will replace them. Of course it's not the whole solution. Capturing all the CO_2 from the transport sector and converting it back into carbon neutral fuel still leaves you with all those coal-fired power plants. But whoever prevails, as a model for partly weaning us off the geopolitical hot potato of hard-to-find, ever-reducing, dangerous-to-drill and easy-to-spill-into-the-ocean reserves of oil (and then replacing them with a

---

*Estimates vary, but pretty much everyone—from environmental pressure groups, to government advisers, to think tanks, to the oil companies themselves—understand that we're approaching a "peak oil" situation at some point in the next generation, a situation where demand will outstrip supply, from which point the gap continues to widen.

carbon neutral alternative) it makes more sense than underpants. Even a certain Australian MP should understand that.

I phone Klaus and ask him if the idea of using his CO_2 as a feedstock for fuels has occurred to him too.

"Yes, I've had some conversations along similar lines with some bio-fuel guys."

"George Church?" I ask.

"No, but please introduce me," he says. A few weeks later I find a quiet moment to drop them both an e-mail, which feels pretty cool.*

I've had an incredible day wandering, it seems, between apocalypse and salvation, and I feel the need for something ordinary and trivial to calm my mind. Colin comes up trumps. Along with a group of his friends, I'm taken on a tour of Manhattan's bars, where we spend a large part of the evening comparing the relative merits of eighties pop icons the Pet Shop Boys and Duran Duran. The latter, I suggest, were more fun and had better songs. Others disagree.

Sometimes, after a day of talking about things that really matter, you need an evening discussing things that don't.

---

*A few months on I get a call from John Ward, Joule Biotechnologies' senior vice president of production, who says the company has had some discussions about "finding a way to make ourselves independent of sources of CO_2" and asks me for more information on Klaus's work. I have become www.match.com for geeks.

10

"I'm a Disruptive Guy.
I Like Disruption."

Most of the shadows of this life are caused by standing in our own sunshine.

—Henry Ward Beecher

started this journey wondering how long I'm going to live. Now I know: about ten minutes if Tracy Wemett has anything to do with.

There's a lot to recommend Tracy. She gives up her free time to mentor disadvantaged teenagers, organize food runs for the homeless and fundraises for an orphanage in Kenya. She's good at her job too—public relations. But her driving is without doubt the worst I've ever experienced. "Nearly there," she says before taking a corner at treacherous speed, while texting. I'm quietly crapping myself. "There" could easily mean the afterlife.

I've never been happier to see a factory building in my life. The sign outside reads "Konarka." Tracy, oblivious to the very real danger my trousers and her sports car's upholstery have just come close to, leaps out of the driving seat. "I love coming here," she tells me as I step tentatively onto the tarmac, relishing its ability to stay perfectly still. "You all right? Do you need the toilet?"

From the outside, this nondescript building in an industrial park just outside New Bedford, Massachusetts, is nothing to write home about. Inside, however, something extraordinary is being made.

———

"This factory used to make photographic film for Polaroid," says Larry Weldon, vice president of manufacturing, who's giving me a guided tour. "That's why Konarka bought it. The process is similar. They bought the building, but they bought the machines too."

Larry accompanies Tracy and me along the production line. Huge spindles feed a continuous sheet of transparent plastic about two meters wide through a series of machines, each depositing thin layers of material onto the film.

"Really it's just a big printer," says Larry.

Like a lot of engineers, Larry is a master of understatement. Because this "big printer" prints rolls of lightweight flexible "organic" solar panels just five millimeters thick that you can wrap around, well, anything.

We've never had an energy crisis. A popular statistic states the Earth receives more energy from the sun in a single hour than we use in an entire year.[1] No, there is no *energy* crisis. What we have is an energy *conversion* crisis. Or more specifically, a *cost* of energy conversion crisis. That's why fossil fuels have done so well. Despite all their annoyances they currently deliver more energy for your buck than other energy sources. That's because nature has taken millions of years to compress dead plants and animals into "hydrocarbons"—materials made of only hydrogen and carbon.

Breaking the chemical bonds between atoms of hydrogen and carbon releases energy in the form of heat and light and is fairly simple (fossil fuels burn easily). It's because fossil fuels have a high density of molecules packed full of bonds ready for breaking that they release more energy when they are burned than, for instance, Twinkies. Running your car on Twinkies would be prohibitively expensive (although do please have a go and let me know how you get on). Another advantage of fossil fuels is they keep their energy nicely locked up until we need it, acting like a natural battery. Fossil fuels are the energy bank account Mother Earth has been paying into, diligently investing billions of years to turn biomass into oil and coal. Unfortunately we're making withdrawals from the account faster than it can be replaced. What nature took billions of years to save, we're spending in a matter of centuries. When it comes to fossil fuels, mankind is like an idiot with a credit card, and at some point the bank is going to cut off our funds. On top of this, fossil fuels pump gigatons of CO_2 into the atmosphere every year, contributing to global warming.

Solar energy as an alternative, therefore, makes intuitive sense. After all, all our power comes from the sun in the first place. It was the sun those ancient plants took their power from, using a process called "photosynthesis." It was those plants that long-dead animals (now, along with their food, turned into long chains of hydrocarbons) fed upon to get their energy. Why not bypass this billion-year waiting game? The sun is waving a huge energy paycheck in our faces every second. The problem is we've not been very good at banking it. Most of it slips away, leaving us to raid Mother Earth's savings account.

Solar power has been a future icon since well before I was born. In his brilliant essay investigating its origins, "The Beautiful Possibility," writer Paul Collins quotes a *Popular Science Monthly* article from the seventies: "Future generations, after the coal-mines have been exhausted, will have recourse to the sun for heat and energy," it announced. "The sun will be the fuel of the future." Given the energy crises of the seventies, this hopeful tone doesn't seem out of place. Except, Collins points out, this article was written in *1876*.[2]

The *Popular Science Monthly* piece was actually a translation of one that first appeared in *Revue des Deux Mondes* (translation: Review of the Two Worlds), a Parisian literary magazine that's been going since 1829 and is still in business today. Back then France was marveling at the work of Augustin Mouchot. In 1869, Mouchot used a parabolic mirror to focus the sun's rays on a copper boiler, creating a solar-powered steam engine,* the same year he published the first book on solar power, *La Chaleur solaire et ses applications industrielles*. At the 1878 Paris Exhibition, Mouchot and his assistant Abel Pifre used a mirror over thirteen feet wide and a twenty-one-gallon boiler to power, ironically, an ice maker—and received the exhibition's Gold Medal for their efforts.[3] The next year, the exhibition was treated to a solar-powered printing press.

Of the burgeoning Industrial Revolution, Mouchot prophetically remarked, "Eventually industry will no longer find in Europe the resources to satisfy its prodigious expansion. Coal will undoubtedly be used up. What will industry do then?" But despite patronage from Napoleon III, Mouchot's invention (unlike the telegraph) never caught on. High costs of manufacture, coupled with the low price of fossil fuels, made it economically unviable—a combination that has dogged the solar energy industry for most of its history. But that's about to change.

Mouchot's "concentrated solar thermal power" method is still in use today but in a modified form and on a much grander scale. In sunny countries across the world, solar power plants with banks of parabolic mirrored "troughs" focus the sun's energy onto thin tubes of hot oil. The oil is used to heat water, which boils and creates steam to drive a turbine. One particularly ambitious modern application of this technology is the proposed Desertec Industrial Initiative, which aims to put huge concentrated solar thermal plants along the North African coast and ship electricity via massive power cables passed under the Mediterranean Sea to provide continental Europe with fifteen percent of its electricity by 2050.[4] Before you think this is idealistic posturing by eco-geeks, the partners in the project include Abengoa Solar (Europe's most successful solar power plant manufacturer), Deutsche Bank, Morgan Stanley and Siemens.[5]

---

*Parabolic surfaces are shaped to collect energy from a distant source and focus it to a common point. Those big radio telescopes use them to collect and focus radio waves from the cosmos. Because the principles of reflection work both ways, you can also use parabolic surfaces to project energy from their focal point outward, which is how car headlights work.

It turns out that our energy industries are staggeringly inefficient. To generate most of our electricity we take fuel out the ground (usually coal), drive it to a power plant, where we burn it in large quantities to create steam to turn turbines. The resultant electricity gets shipped along miles of cable to the end user. Coal plants lose over fifty percent of the energy generated as waste heat[6] and up to ten percent of what's left can be lost in transmission down the wires (the figure was 6.5 percent in the U.S. in 2007).[7] I find this staggering. It's what you might call a lost opportunity, like being offered a date with Elle Macpherson and deciding to go to the cinema by yourself instead. Energy conversion crisis indeed.

For our liquid fuels we drill for oil in the few locations we can find it (less than fifteen hundred drilling locations account for ninety percent of the world's known oil reserves) then ship this to refineries, where it's processed before distribution to retailers. If you want any, you have to go and collect it (e.g., from a gas station) or arrange a delivery. You then burn it, but miss out on most of the energy produced, losing up to eighty-five percent.[8] That's like being offered a date with Elle Macpherson and choosing to go on one with Shane MacGowan.

While this all sounds horribly inefficient and inconvenient, it's still the cheapest way we've found to answer our energy needs. Solar power might sound like an obvious option but technologies for turning sunlight directly into energy have suffered from inefficiencies even greater than those we currently endure.

At the same time, there's a clear case for improving the efficiency of the existing system rather than switching to new energy sources. While in California, I'd popped in to see Dan Reicher, Google's head of environmental projects. Dan argued that concentrating on energy efficiency is the best thing we can do for the planet in the short term. "I apologize that this not the exciting stuff and therefore may not make it into your book, but the low-hanging fruit is doing more with less; energy efficiency across the entire economy."

Part of that solution will be the "Internet of things" I talked about with Vint Cerf. "We're starting to see smart appliances entering the market," Dan said. "The more they can talk to each other and your electricity supply about how much they're consuming, the more they'll be able to coordinate to use less electricity. Do I really care whether my dishwasher runs at six o'clock when it's a hundred degrees out and the electricity system is browning out because everyone's got their air-conditioning on, or whether it runs at three in the morning at half the cost with a lot less impact? I probably don't care as long as my dishes are washed next morning. We've developed the beginnings of some software where the grid can talk to various appliances in your house. Instead of turning on the next big power plant, the grid may say 'we want you hundred thousand refrigerators not to go into defrost cycle for the next thirty minutes.'"

The argument is that energy efficiency is the quickest and cheapest way to reduce the CO_2 load in the atmosphere. Short term, Dan is probably right, but making our use of fossil fuels more efficient doesn't change the fact that we'll still be using them (and using them up). We'll still be pouring CO_2 into the sky, even if we are doing it more slowly.

"This is one reason Google is developing new mirror technologies to help bring the price of solar thermal plants down," says Reicher. "We're looking at new and cheaper materials both for the reflective surface and the substrate that the mirrors are mounted on and hope to halve the cost of heliostats—the fields of mirrors that track the sun. Ideally we'd like to cut the cost by a factor of three or four."

I'm convinced that concentrated solar thermal energy has a large part to play in our future, but I've come to New Bedford because what's happening here is part of the other half of the solar revolution—and it could see a radical shift in our relationship with energy.

Traditional silicon solar cells ("crystalline silicon cells") work because light energy hitting a silicon wafer messes with it, knocking negatively charged electrons off atoms. Each cell consists of two wafers, one placed on top of the other—one with added boron and one with added phosphorus. Mixing boron and phosphorus into silicon is called "doping" and gives the wafers different electrical properties. The phosphorus-doped "negative" wafer faces the sun, and when an electron gets dislodged in this part of the cell it will head off in the direction of the boron-doped "positive" wafer underneath, setting off a chain reaction of "charge passing" between atoms and creating an electrical current.

Chris Goodall, author of *Ten Technologies to Save the Planet*, recalls installing twelve thousand pounds' worth of crystalline silicon solar panels and associated electronics in his house in Oxford. If he'd sold (rather than used) the power they had generated over a four-year period, "I would have banked a check for three hundred pounds," he writes. That's like being offered a date with Elle Macpherson and deciding to kill yourself. If he'd bought a more efficient car, he'd probably have done more for the planet.* Part of the reason silicon cells have been expensive is that the process of manufacture is complex and requires a huge amount of energy, which ironically gives them a large carbon footprint (the "doping" process, for instance, requires the wafers to be heated to slightly below the melting point of silicon).

In recent years, "thin-film" solar cell technologies have emerged that use much slimmer wafers of silicon or combinations of materials with catchy names like "cadmium telluride" and "copper indium gallium diselenide." These cells can be made much thinner and lighter and are often cheaper than their crystalline silicon counterparts, which means even though they generate less electricity per square foot, the cost of the energy produced can compare favorably. One (perhaps overstated) concern is that tellurium and indium are some of the rarest elements on the planet, raising worries that, ironically, these "renewable" technologies will be hampered by the shortage of a natural resource. Another worry is that cadmium is one of the most toxic elements known to man.

---

*Admittedly, Chris's system was bought nearly ten years ago, so the figures would be better with a more up-to-date installation.

Back in the factory, we've paused at a machine where the impossibly thin power-generating layer is being printed onto a roll of plastic passing through it. "We don't use silicon, tellurium, indium, gallium or any of that," says Larry.

"What do you use?" I ask.

"Plastic ink."

Konarka's technology uses molecules of "organic conductive polymers," jointly discovered by the company's cofounder Alan Heeger, along with Alan MacDiarmid and Hideki Shirakawa (a discovery for which the three shared the 2000 Nobel Prize for Chemistry). The "organic" label refers to the fact that the polymers in question are partly carbon. Because carbon is such a versatile atom (it will form all sorts of chemical bonds) it's the one nature prefers as the key building block for living things, which is why aliens in sci-fi movies of the seventies like to refer to humans as "carbon-based life forms" (it's the first thing their scanners pick up). It's not just us though. All plants and animals are carbon based. This is why chemistry involving carbon is often referred to as "organic" chemistry.

Like the two types of doped silicon used in traditional solar cells, conductive polymer molecules can be either "positive" or "negative." Like the phosphorus and boron-doped wafers in silicon cells, one set of polymer molecules gives up electrons and the other collects them. However, unlike the silicon wafers stacked neatly on top of each other, all these molecules are mixed up together in an inexpensive ink. The good news is that you can literally print solar cells. The bad news is that because the positive and negative molecules are all jumbled together, some of those electrons don't make it very far, getting captured back almost as soon as they've been released. So the material currently generates low levels of power measured in proportion to its surface area. Konarka's material, though, is much cheaper than silicon cells and can work in low light (and therefore for longer periods of time). In fact, before we'd started the tour, Larry had shown me a sample happily generating electricity under

the fluorescent tubes dimly lighting the office where Tracy had introduced us.

How do you know which solar cell is best? Sadly, comparing the cost of
energy produced by different solar technologies is difficult. Manufacturers
like to quote "efficiency ratings," but this doesn't help much. If a solar cell
is quoted as "ten percent efficient" it doesn't mean it will always convert
ten percent of the sunlight hitting it into electricity. A cell might be "ten
percent efficient" in bright sunlight but generate nothing in lower light
(the light hitting silicon cells, for instance, has to pass a certain threshold
before they "switch on"). Some cells don't make use of light coming in at
shallow angles (early in the day or as the sun sets). On top of that, once silicon cells heat up past a particular temperature, their efficiency drops. Certain thin-film technologies by contrast show *increased* efficiency at higher
temperatures. Taken together this means that a lower efficiency cell might
deliver *more* electricity over time. So efficiency ratings aren't very useful
for telling you how much electricity you'll get out of your cells over their
lifetime (between twenty and thirty years for silicon cells).

Cost per kilowatt-hour is a better, if admittedly imperfect measure—
providing a way not only to compare different solar technologies but
all energy-generating options, from a coal-fired power plant to a wired-
up hamster wheel. Simply take the lifetime costs associated with each
(including construction and running costs) and divide this by the lifetime
total of usable watts of energy it delivered in all of the hours it was generating power. This calculation explains why hamsters never got into the
utilities business.

Sadly, though, you soon run into problems using this method too.
Delve just a little way into published comparisons of energy generation
technologies and you'll find yourself knee-deep in assumptions and caveats. For example, any comparison is victim to fuel price fluctuations over
the time period in which it was made. Beyond that, solar energy is literally
dependent on the weather. The cost per kilowatt-hour over the lifetime

of identical installations will be different if one is in Damascus and the other is in Dartford (or even one in Dartford and one in Dagenham).

We're near the end of the tour, and Larry is explaining one of the reasons Konarka's material (they call it "Power Plastic") could be cheap to produce on a large scale.

"We can retrofit old photographic film plants," he says, and although he won't confirm the figures, it's clear Konarka bought the building we're now standing in at favorable price. "Polaroid wound up production in 2007. I was out of a job." But not for long. Konarka bought the plant and rehired Larry and thirteen of his staff to refit the operation. I'm looking at the fruits of their labors. Rolling past me are yards of film, all of it being turned into power-generating material. Its "efficiency" might be low (currently between three and four percent depending on the batch) but this plant looks more like a newspaper print-works than a solar cell factory. Konarka claims that in a single shift they can make a square mile of material.

As we walk to the exit, I'm amused by a machine that proudly displays its name as "Butt Splicer B." Tracy and I share a joke that before Polaroid was here this factory manufactured asses and this was the machine that put the crack in. Its actual function is far more mundane—simply butting two films up to each other for joining together, but I'm adamant that whoever named this knew what they were doing, and to be honest, who can blame them? Given the options of "dual-film joiner" or "butt splicer" I know which I'd choose.

I thank Larry for taking the time to show me around and begin to make my way out of the factory with some trepidation. I know the only way I'm getting back to my meeting with Konarka's CEO Rick Hess is with Tracy at the wheel.

Rick Hess is what I would call "very CEO." He has the confident, relaxed air of a man who probably doesn't have to worry too much about his pension. His face is strong with big features: strong jaw, confident nose, eyes that meet your gaze full on. He doesn't so much sit in the chair opposite me as own it. It's not arrogance, more purebred entrepreneur-grown assuredness. Rick builds companies, sometimes sells them, and clearly revels in the corporate environment. He's also generous about his competitors. "I'm a total believer that there are niches for everyone. Thin films certainly won't put silicon out of business, but they can be used in places where silicon can't." He's also candid about the current limitations of his own firm's technology, notably its life span.

One of the major attractions of Konarka's technology is its flexibility, allowing it to be easily wrapped around anything from buildings to back-packs. Konarka even own the patent to solar thread.* However, because the conductive polymers they use are "organic," they quickly react with oxygen and moisture in the air. To keep flexibility but protect the power-generating material from these damaging reactions, Konarka needs to wrap it in a flexible airtight coating. "The material we package lets mois-ture and oxygen in after three to five years," admits Rick. "We've tested ones that will give our technology a lifetime of over ten years but they're too expensive at the moment. We expect that to change."

Another advantage is that Konarka's films can be made thin enough to see through, which means they can be used to coat windows and turn them into power generators. It's not a flexible application, but because the material can be put in the vacuum between double glazing it enjoys a lifetime of between fifteen and twenty years, allowing a building to gen-erate power as long as there's light hitting the windows. It's one example of something your electricity supplier probably doesn't want you to think about—going "off grid."

"Phones went wireless and Internet went wireless and the only thing that's left that you have to find a wire for is power," says Rick. "You think about the developing world and what happened with communications.

---

*The problem with that, says Rick, isn't generating the power, it's hooking up enough of the thread so the electricity can be put to use.

They skipped wires and went straight to wireless. I think for power they're going to do the same thing."

A few months after my visit, the company began working with one of their customers to create solar-powered lanterns for developing countries, devices which store up charge in the day and can be used at night for study or work, as well as improving personal safety. "Having light really does help a village and its people move to the next level of development," says Rick. He's right. The village of Rema, a hundred and fifty miles north of Ethiopia's capital, Addis Ababa, has installed solar power. "Our kids can do their homework at night now, because there is light. They are very happy," says resident Elfenesh Tefera. Meanwhile the local bar can operate for longer hours and a solar fridge ensures the beer is always cold and the staff and customers no longer have to struggle with smoke from gas lamps. Since solar came to the village, it's booming. It's the place to be with an influx of newcomers.[9]

"Konarka cofounder Dr. Sukant Tripathy's goal in starting the company was to ensure that developing countries were able to get power," Rick explains. (The company was named in honor of the late Tripathy's favorite place, the Konark sun temple in Orissa, India.)

I ask Rick if he'd heard about a fuss a few months before which erupted when Xcel Energy, Colorado's biggest utility company, attempted to levy an extra charge on customers with domestic solar panels as a way to help pay for an expansion of the grid in the state.[10]

Rick laughs quietly. "The utilities are doing the halfway model. They want solar power but they want to deliver it over the wires. There will be a place for that. Depending on where you live, it might make sense to buy your electricity from somewhere sunnier, where they can produce it cheaper."

I get a strong impression Rick is playing the professional "be nice" card. Others are more direct. Jim Harvey of the Alliance for Responsible Energy Policy is steadfastly against the idea of solar power being owned by the utilities. "The utilities are more interested in protecting their stranglehold on the power grid and preserving their century-old business model than they are in producing clean electricity," he says.[11]

"Are you talking to utilities?" I ask Rick.

"No."

"Because . . . ?"

Rick doesn't answer directly, but tells a story. It's about a major telephone company sponsoring research into cell phones several decades earlier. "In their grand wisdom, five years into the program they said, 'This makes no sense! Nobody wants to carry their telephone around with them! We have wires all over the United States and people just go and pick up the phone and call. Why would they ever *carry* their phone with them? This technology is going nowhere.'"

His inference is obvious. The days of an electricity system dominated by centrally generated power are coming to an end, that solar could do for electricity what synthetic biofuels might achieve for gasoline: create a world where our energy sources are hyper-local. While I was at MIT talking to roboticist Cynthia Breazeal, I'd taken the opportunity to meet with Bill Mitchell, head of the Smart Cities Group there. "We're clearly going to see a movement away from old-fashioned centralized electrical grids to a much more decentralized system that looks a lot like the Internet," he'd said. It's a nice idea, but there's still a huge distance to cover. Solar power remains expensive and there's a lot of inertia in our existing infrastructure. Rick agrees.

"I won't say when it'll happen, that's for the futurists to think about. We're just getting on with making it happen. You just can't take the grid to everyone. It's not feasible." He's right. Two billion people live without access to reliable electricity supplies.[12] In Ethiopia, for example, eighty percent of the population is rural, of which just one percent has access to electricity. A truly national grid would be prohibitively expensive, which is why the country is experimenting with rural off-grid solar, as are Nepal, Sri Lanka, India, Vietnam, Ecuador, Tanzania, Indonesia, Kenya, Brazil, Ghana, China . . . well, you get the gist.*

---

*China is also forging ahead with ambitious grid-solar projects. The country is building a thin-film solar power plant in the Mongolian desert that's larger than Manhattan. When completed it should power three million homes.

A few months after I meet Tracy, Larry and Rick, something catches my eye. The town of Fowler, Colorado (the same state where Xcel got into trouble over its proposed solar power surcharge), announces plans to go off-grid with a combination of solar power, homegrown biofuel and gas derived from manure. "The primary goal is to stabilize utility costs and then to reduce them," says Wayne Snider, the town manager, "but our ultimate goal is to become our own utility."[13] It's a hint of the hyper-local energy revolution Rick is talking about, in the heart of America.

"What about the storage problem?" I ask Rick. "Solar only generates when the sun's shining."

"Of course you need batteries to store excess energy generated in the day."

One storage option is using the material-pimping possibilities of nano-technology. As I found out in California, nanotechnology is infiltrating everything with the promise of driving down costs and improving per-formance. The energy industry is no different. (Indeed, Konarka's con-ductive polymers are "organic" nanotechnology.) Work at the University of Maryland on "electrostatic nanocapacitors" aims to increase battery storage by a factor of ten, and they're already thinking about integrating the results into solar cells.[14] MIT researchers are working on nanostruc-tures that will hold charge indefinitely.* [15]

Rick's solution? A *printable* polymer lithium battery. "Our technology isn't limited to solar cells. You can make electrical components out of polymers too. So. for example. you could make a product with three lay-ers: solar panel, polymer battery and organic LED light, all printed. A lot of hot countries need shade, so you make your shade out of this material. It keeps you out of the sun in the day and allows you to see at night." This is the other reason I came to New Bedford, to get a glimpse at a wider

---

*Nanotechnology is also applying itself to improving the grid. Researchers at Rice University in Houston are working on ways to use super-conducting carbon nanotubes to do away with energy losses in transmission. The project, started in 2001 by Eric Drexler's former sparring partner, Rich-ard Smalley, has created carbon nanotubes hundreds of meters long but only fifty thousandths of a millimeter thick. As yet, however, there is no way to mass-produce the right kind of nanotubes, but they're working on it.

revolution. Heeger's conducting polymers could change the way we manufacture electronic devices in the same way Gutenberg's printing press changed the way we made books. This might explain why universities around the globe are setting up organic electronics laboratories at a rate of knots, many focusing on organic solar cells. A quick Internet search pulls up more than seventy in countries as diverse as America, Germany, Japan, India, China, Morocco, Australia, Ethiopia, France, Canada, the Czech Republic, France, Greece and Slovakia.

In an example of just how fast the technology can be made and deployed, when the Haiti earthquake struck the year after I visited New Bedford, Konarka employees used their weekend to build battery-charging units for Boston-based Partners In Health (a nonprofit dedicated to preferential health treatments to the poor), delivering them within twenty-four hours. The company tells me PIH used the devices to provide power to places where there was no grid electricity so doctors and medical staff could work around the clock.

Each year the U.S. Energy Information Administration publishes its International Energy Outlook, presenting several possible future scenarios for energy consumption, including one where laws and policies relating to energy production stay the same as today. Like all energy predictions, it's based on assumptions and figures that you could argue with (and people do) but they're one of the few organizations who've had a crack at looking at the whole world (and drawing on my own experience of trawling through thousands of pages of energy statistics, I believe that alone merits praise).

It makes for grim reading if you're worried about man-made global warming. The 2010 report estimates that by 2035, world energy use will rise by forty-nine percent from 2007 levels. While the use of renewable sources grow, so does the use of fossil fuels, with the result that CO_2 emissions go up forty-three percent.[16]

"Can solar scale up?" I ask Rick.

"Not all of the industry can, no. That's where a roll-to-roll process comes in. We can make solar material as fast as the spindles run and there's a lot of printing facilities in the world sitting empty. When the time comes we can scale up fast."

While I know Rick is giving me the company line, there's already a precedent being set. People often quote Konarka in the same sentence as California-based Nanosolar, which also uses a roll-to-roll thin-film manufacturing process, coating their material with "nanoparticle ink." Nanosolar forgoes flexibility, mounting their material on a rigid substrate, but still claim they can produce a solar panel seventy-six inches long by forty-one inches wide[17] with eleven percent efficiency[18] every ten seconds.[19] Unlike Konarka, Nanosolar is concentrating on utility scale implementations. A few days before I meet Rick, the company announces orders totaling $4.1 billion from a number of solar power plant builders.[20]

"Our scientists have goals this year to achieve two to three times the efficiency of our current commercial products," says Rick. "With biochemistry you can engineer molecules to do incredible things. For us it's a matter of designing polymers that both absorb more light and more frequencies of light," says Rick. "It's some pretty interesting chemistry but we're onto it."

Again, it's the company line but I'm inclined to believe him. Not because of his winning smile and confident drawl but because I've seen what George Church is doing at Harvard in the field of synthetic biology. If Church can engineer living cells that make fuel, it's not hard to imagine that experts in synthetic chemistry like Alan Heeger can continue to push the boundaries of their science to make organic materials more efficient at generating electricity. "We can also make materials that generate electricity from infrared heat radiation," says Rick. "So one of our patents involves a product that generates electricity from sunlight on its top surface but electricity from infrared on its underneath. The idea is you put them on the roof of buildings and use waste heat coming up as well as the sunlight coming down to generate power."

It's on the promise of advances like this that techno-optimists claim that widespread "grid parity" (the point at which the price you pay for energy is the same whether it's been made by solar or fossil fuel) is but a hop, skip and a jump away. Others are less convinced, like self-confessed solar skeptic Dan Lewis, research director at the UK's oldest economic think tank, the Economic Research Council. Dan believes that "grid parity may become a reality in some sunny, major parts of the world by 2020."[21] A particular problem for Konarka is a classic "chicken and egg" situation. Its "Power Plastic" material will start to get really cheap when production scales up to industrial levels, but until it *gets cheap* there won't be the demand for that level of production. (This is probably the real reason Rick isn't talking to utilities—right now his product is too expensive per watt to be of any interest to them.)

Rick announces he has to leave for a meeting but before he goes I want to ask him a final question. Rick is a guy who likes to make money. He's not only in solar for the good of the planet.

"Why are you doing this job?"

"I'm a disruptive guy. I like disruption. And solar is going to disrupt the energy market." He pauses. "But it does feel good to work on something that's a really good cause that everyone identifies with. My son is interested in what I'm doing. He never was before in his life! Dad's doing something that's cool!"

We say goodbye and Rick gives me one of those impossibly firm corporate handshakes that seems to say, "Let's be friends, but know I will kill you if called upon." Tracy turns to me.

"Need a lift?" she asks.

"You know what? It's sunny," I say. "I'll walk."

Clean energy fans and concerned environmentalists shouldn't bank on rising fossil fuel prices to stimulate demand for renewable sources. The fossil fuel industry has a long history of finding new supplies to keep prices generally steady, even while they're losing millions of barrels elsewhere in ecological disasters. While no one denies that fossil fuels will run out someday, this won't be what switches us to renewable sources in the time scale needed to halt our prodigious output of CO_2. These sources have to get cheaper, and fast. But the level of innovation in the solar industry is extraordinary. There is no equivalent, for instance, of roll-to-roll manufacture in wind power. Thin-film manufacturers are springing up quicker than doomsayers at an England soccer game. Every week seems to bring an announcement from one firm or another claiming their cells are pushing efficiencies up and costs down. Not all can succeed but some will, and I suspect it'll be the ones who exhibit the smartest use of nanotechnology ("organic" or otherwise) that will prevail.

Depending on which report you read and which year it was written in, solar energy is either the best-kept secret in the energy sector, showing exponential growth that will see it replace fossil fuels in twenty years, or a curious and insignificant sideshow. A more coherent, if counterintuitive, analysis is that it's both. Solarbuzz, a solar industry research and consultancy firm, estimates demand for solar energy "has grown at about 30% per annum over the past 15 years."[22] But these figures come from a world where "grid parity" has yet to be achieved. What happens when solar starts to get cheaper than the alternatives? Given that even solar pessimists believe grid parity will be reached in some sunny countries within the decade, it's not unreasonable to speculate that another decade on, solar will get much cheaper in those places (possibly stimulating a radical shift to local off-grid energy generation) and achieve grid parity in many other locations. Would we then see solar demand growing not by thirty percent per annum, but fifty percent? An annual growth rate like that would see solar quickly match all our needs, turning a sideshow today into the main feature tomorrow. It's instructive to look at the growth rate in Japan—one of the countries pretty much all analysts think will see grid parity in the near future. For 2009, Solarbuzz quotes a growth rate of 109 percent.[23]

Here comes the sun and I say, "It's all right." (Are you humming the fiddly guitar bit?)

11

"The Black Phantom"

> The best time to plant a tree was twenty years ago. The next best time is now.
>
> —*Chinese proverb*

I've come to the other side of the world. It's different here.

Presenting myself to airport immigration, I'm expecting a familiar attitude of suspicion and disdain. I generally come away from national borders with the strong impression that most immigration officers are hired precisely because they have an all-consuming dislike of other people. I get the feeling the highlight of their working day is the chance to carry out an invasive cavity search.

"What's the purpose of your visit?" asks the fresh-faced man behind the passport counter.

"I'm researching a book."

"Oh really? What's it about?" comes the cheery response.

Is this a trick, the jolly demeanor?

"It's about the future," I say carefully.

"Oh great! I've got some ideas about *that.*"

I'm apprehensive. Is this his way of telling me *my* immediate future involves a rubber-gloved official poking something up my bottom in search of illegal drugs?

"I think solar is going to be really important," he says.

I'm taken aback and without thinking reply, "It is!" before showing him the bag I'm carrying. Given to me by the good people of Konarka, it incorporates one of their organic solar panels and has been happily charging my iPhone since I left New Bedford. Before I know it, we're in a conversation about the role of technology in solving the world's problems. (He imagines a future of self-driving magnetic cars.) I look behind me expecting to see a queue of knackered and exasperated fellow passengers waiting to go home. Instead, they're all streaming through other channels exchanging friendly greetings with immigration staff. Have I entered an alternate reality? A border that deals with arrivals quickly and politely? One that actually *welcomes* you to the country? I'm already enjoying New Zealand and I haven't left the airport.

"Who are you seeing?" asks the guard.

"A lady called Vicki Buck."

"Vicki's great," he says, and I'm taken aback a second time.

There again, it's not such a huge surprise. Vicki was the mayor of this city (Christchurch) for nine years, popular for her initiatives in tackling unemployment, invigorating the town's cultural life and addressing housing and social care issues. Today she's leading a charge for the renewable energy industry in New Zealand that could have international repercussions—initiatives that are undoubtedly good for the planet whether you worry about climate change or not.

The guard takes my blog address and promises to buy the book. "Don't forget to mention those self-driving cars!" he says.

Despite being tired after nearly a day crammed into a procession of economy-class airplane seats I practically bounce out of the terminal. I have a strong feeling I'm going to love this country.

Vicki meets me at my hotel the next morning to drive us to breakfast. Although to say one "meets" Vicki is something of an understatement. Vicki meets me in the same way a tornado "meets" the air. You are instantly swept up, and it's invigorating.

The first thing you notice is she cannot stop laughing. She laughs at everything. It's not the laughter that comes from a good joke, more an all-consuming *joie de vivre*. Anything can raise a smile, animating her broad features into a mixture of delight and mischief. In her early fifties, Vicki has the energy of someone in their early twenties, but coupled with decades of practical experience that, if anything, has emboldened her rather than beaten her down. Hard-won knowledge is married with an unbridled belief in the possible. In the week I spend here, I see Vicki wield authority without being authoritarian, command respect without demanding it and motivate action without telling anyone what to do.

In 1998, Vicki quit politics to concentrate on education, helping to set up two schools in the city and in which she is still deeply involved. With these established, she began committing her almost inexhaustible energies to tackling the climate change issue. While she knows the state has an important role to play in dealing with global warming, her overall view is, "if we wait for our governments to sort it out we're probably all fucked." Oh yes, she has a reputation for straight-talking too. This might explain why she preferred to stand for office as an independent rather than as a candidate for a political party. She's not wedded to any ideology other than the potential wrapped up in every individual. "I like the 'just do it' approach," she tells me. "Just go and make it happen."

"Have you heard this song?" she asks, dialing up a CD on the car hi-fi.

The opening bars of a jaunty pop number issue forth. Vicki is tapping the steering wheel in time to the music, the sun roof is down and for three minutes we listen to Kiwi songsmith Tim Finn's "Couldn't Be Done" inviting us to prove all doubters wrong. It could be Vicki's theme tune. I get the impression it's on strong rotation in this vehicle. "Isn't it great?!" she exclaims.

"I think you can achieve a lot more if you're enjoying yourself," she tells me as we order breakfast, possibly explaining one reason why Vicki accomplishes so much. The list is frankly frightening. She runs through a small selection of her current interests: she's a director of

New Zealand Wind Farms, of biofuel company Aquaflow Bionomic Corporation, the climate action website Celsias, and charcoal manufacturer Carbonscape.*

It's this last company that has snagged my interest and the main reason for my visit to New Zealand. Charcoal has undergone a radical makeover in recent years. Creating it (and then burying it) is seen by many as one of the most promising ways to mitigate global warming. James Lovelock, who thinks we're heading for climate Armageddon, has even suggested, "There is one way we could save ourselves and that is through the massive burial of charcoal."[1] In this context as a planet saver, humble charcoal has been given a flashy new brand name: biochar.

The biggest contributors to CO_2 in the atmosphere are plants and animals (both on land and in the sea). In fact, they dwarf human emissions. "The biosphere pumps out 550 gigatonnes of carbon yearly; we put in only 30 gigatonnes," says Lovelock.[2] Skeptics often point to this difference and say, "See! Global warming has nothing to do with us!" Sadly, though, it does.

While it's true that the lands and oceans burp out massive tonnages of CO_2 they also take it back. In fact, it's a finely balanced system—what gets emitted gets absorbed again. This is called the "carbon cycle." Through the miracle of photosynthesis, plants play their part by taking CO_2 from the atmosphere and turning it into "biomass." Nearly sixty percent of a plant's mass is carbon taken from the atmosphere.[3] Most of that carbon is returned to the air when a plant dies—the vegetation rots and the carbon is released skyward. It's a natural cycle. Unfortunately, our use of fossil fuels has taken carbon locked deep underground (after slowly being buried and compressed over millions of years) and released it back into the atmosphere. The good news is that nature *is* compensating for

---

*You'll also find Vicki working on forestry projects, talking at international conferences on climate change, negotiating with electricity providers on how to deliver domestic solar systems and always, always laughing.

these extra emissions, but not completely. The land and seas are taking up some of the extra, but not all of it. Half of it is remaining in the sky. The result is a rise in atmospheric CO_2 that contributes to global warming and, as I found out in New York, it doesn't take much extra CO_2 to have a big effect.

At the heart of the great charcoal/biochar resurgence is a process called "pyrolysis"—a way of burning biomass in a "low oxygen" environment that creates a stable nonbiodegradable form of carbon: charcoal. Unlike a normal open fire which turns what you burn into ash and CO_2, pyrolysis "part combusts" it in an oxygen-starved enclosure, turning roughly half of it into charcoal (carbon) and the other half into wood gas and oil that can be used as fuel. ("Wood gas" sounds like something out of a Tolkien novel but it has a long history as an energy source. In fact, vehicles powered by it were common during World War II when fossil fuels were rationed. Volkswagen and Mercedes had production models.[4])

Charcoal is locked-up carbon that, like coal, won't be released unless it's burned. It doesn't rot like the biomass it used to be before it was pyrolized. There's a neat symmetry to this. It's the release of carbon trapped in ancient buried biomass that's causing increases in CO_2 levels in the atmosphere. Here, contemporary biomass is being used to retrap that carbon and stick it back in the ground.

Cornell University's soil scientist Johannes Lehmann (who shares the same campus as robot maker Hod Lipson) believes the potential benefits could be huge. Of the sixty gigatons of carbon taken out of the sky by plants, he estimates ten percent ends up as waste in agriculture or forestry. Corn stalks, rice stalks, branch and leaf litter (as well as animal poo) are all potential food for pyrolysis, "enough to halt the increase and actually decrease the level of atmospheric carbon by 0.7 gigatons a year," he writes. "Clearly, the potential contribution of biochar technology is large, perhaps large enough to mitigate climate change alone."[5] James Lovelock says this approach could "start shifting really hefty quantities of carbon out of the system and pull the CO_2 down quite fast."[6] It's worth pondering that point. Lehmann is saying that enough waste biomass alone could, if turned into charcoal, not only offset all our carbon emissions but also start to bring CO_2 in the atmosphere *down*, reversing the effects of global warming.

It's important to note that Lehmann is one of the more optimistic voices when it comes to biochar. Others quote lower estimates for the amount of waste biomass available. But there's no denying billions of tons of it are produced each year. Beyond this, however, there is the issue of convincing farmers and foresters to become pyrolysis converts. This is where companies like Carbonscape come in. Before I find out more about them, however, Vicki wants to introduce me to her latest obsession.

I'm looking at a system of municipal wastewater ponds on the outskirts of Blenheim, a small town in the wine-growing regions on the north coast of New Zealand's south island.

"I bring you to all the best places!" laughs Vicki.

Two nondescript blue shipping containers sit beside the ponds, proudly bearing an Aquaflow logo. A man wearing Wellington boots, jeans and a T-shirt with the image of a goat's skull on it emerges from one of them. I instantly think "engineer" and I'm right. This is Mark Vinsen, who explains to me what's going on.

"Water from the pond comes in," he says, pointing to a pipe. "Then we harvest the algae out of it."

"Algae are very cool," enthuses Vicki.

"Those algae have been feeding on contaminants in the water. By removing them, we're most of the way to turning wastewater into fresh clean water," Mark continues. "We put the water through a few more steps but the algae do the initial hard work."

"Clean water is the major part of the business," says Vicki.

Mark takes me to the end of the second blue container, where a green sludge that looks like a cross between pesto and snot is dropping off a roller. "This is algae paste. We process it to create something we call 'biocrude' that can be made into fuels." (Aquaflow has manufactured synthetic components of aircraft fuel, for instance.[7])

"That's the other part of the business," says Vicki.

I'm glad I've been distracted from charcoal for a minute. After learning about the potential of organism-made biofuels from George Church, it's nice to see the science made real. This isn't just an idea in a laboratory, here's a blue box in a field making the stuff.

"What kind of algae do you use?" I ask. I'm expecting to hear about a specifically engineered algal strain that Aquaflow owns the patent for.

"Whatever's in the pond," say Mark. "That's the beauty of it. There are thousands of varieties of algae and they vary depending on the time of year."

"You work with what nature already puts in the pond? You're algae sluts?"

Vicki laughs. "Yeah, when it comes to algae, we'll go with anyone! But what that means is that we can turn any municipal waste pond into a manufacturer of cheap clean water and biofuel."

It's a pretty cool idea and one that will no doubt find its place. In fact, a few months after my visit to Blenheim, the company announces a partnership with the United States Gas Technology Institute to "demonstrate the conversion of algae biomass directly to gasoline and diesel fuel."[8]

We bid Mark farewell and drive off with Tim Finn keeping us company.

"How did you get into this?" I ask Vicki.

She laughs. "I know! Algae? I mean I knew *nothing* about algae. But now? Now I love them. It's amazing what they can do. Seriously, I'm in *love* with algae. I'm a total algae bore!"

Twenty minutes later I'm in a shed on an industrial estate staring at the machine that's brought me halfway around the planet. It looks like it was designed on a drunken night out after a special effects conference. I imagine a tussle between the *Harry Potter* team and the *Star Wars* crew.

"It should have a big cauldron, right in the middle!"

"Well, if you're going to have that, we want a bank of computers on the side."

"Okay, but it has to be *black*!"

It even has a fantasy sci-fi name.

"Behold 'The Black Phantom,'" says Vicki.

This is the jewel in Carbonscape's crown: a black metal edifice that wouldn't look out of place in a Tim Burton movie. About eight feet high, it's dominated by a huge cauldronlike pot, large enough to contain a man. In fact, as I approach, a man in dark blue overalls holding a screwdriver pops his head up over the rim.

"Hi," he says. "I'm Greg."

Standing next to me is one of Vicki's codirectors at Carbonscape, Tim Langley—a man who seems to mix the countenance of a lumberjack and a monk in a single package. There's something about Tim that exudes complete acceptance of life, whatever it brings. It doesn't take much for his face to smile, and when one comes it's not really like any effort is needed to make it, more that a light has been shone onto a grin that was there all along. I ask him to explain the Phantom to me.

"Well, essentially it's a big microwave oven," he says.

"I see."

"Inspired by a potato."

"Of course."

The machine is the idea of paleoclimatologist Chris Turney, professor of physical geography at Exeter University, UK. As a teenager he baked a potato for forty minutes in the kitchen microwave creating a mass of potato-sized charcoal. "Years later I was thinking about a way of turning carbon into charcoal," he told *The Times*. "So we came up with Carbonscape."[9]

"The problem with standard methods of pyrolysis is it's hard to control the type of material you get," says Tim. "Different raw materials create charcoal, gas and oil with different characteristics, and so do the conditions you burn it under, the mix of oxygen, the temperature and so on. To get a consistent result you've got to cook it right. That's where the Phantom comes in: consistent results." Those "consistent results" make Carbonscape of huge interest to many biochar advocates and are the reason I'm here.

Certain types of charcoal added to soil can radically improve agricultural yields. In 2009, Dr. Paul Blackwell of the Department of Agriculture and Food in Western Australia and colleagues reviewed soil charcoal studies from around the globe and were able to quote regular crop yield improvements of over thirty percent and as high as fifty percent. But they were careful to point out that "variability is high and it is not yet clear under what soil and climatic conditions and plant species high or low yields can be expected."[10] The right mix can send yields rocketing, the wrong type can reduce them. A study by Imperial College London found that using soil charcoal improved UK barley yields substantially but *only* when large amounts of artificial fertilizer were also applied. For soils without added fertilizer, increasing the amount of charcoal slightly *reduced* the crop yield.[11] Part of the problem is that no one is quite sure why charcoal tends to have a beneficial effect on soil fertility. The general feeling is that it is more effective at retaining nutrients in the soil (and keeping them available to plants) than other organic matter.[12]

"The more we understand the relationship between charcoal and soils, the more we hope to give farmers or foresters the ability to 'dial up' exactly the right charcoal to increase their yields," says Tim. The advantage for the planet is this will not only increase soil fertility, it will take carbon out of circulation. It's a possible win-win scenario.

"Actually, it's a potential win-win-win situation," says Vicki. "There's the oil and gas they can get out of the process too, creating fuels out of waste."

I'm reminded of a favorite childhood TV program, *The Wombles*, that chronicled the life of imaginary burrowing creatures living underneath Wimbledon Common in London. These furry creatures with huge pointy noses recycled rubbish left by humans into useful products. (I remember one episode in which some old string and conkers were refashioned into an abacus to help one of the Wombles with his math.) "Make good use of bad rubbish" was their motto.

The development of the Phantom has taken a long time and seen a few hair-raising incidents. I'm not surprised. After all, it is basically a big experimental cooking pot with a massive industrial microwave attached to it.

"During one of the early experiments the cauldron's lid blew clean off," says Greg the blue-suited engineer as he climbs down from the pot.

Along with his colleague Forrest, he shows me around the Phantom. Along the back wall of the shed in which the machine lives is a bench strewn with apparatus and tools. At one end is an ordinary kitchen microwave bearing the brandname "Zip." Propped up at a jaunty angle with wires and sensors coming out of it at various angles, it looks like it's come out of a firefight. Several blackened areas of plastic bear witness to its rude treatment.

"We test out basic ideas on this," says Forrest. There's a kind of childlike glee in his eye that I'd also seen when Greg had mentioned the lid-blowing incident. I get the impression he might have been a kid who liked to set fire to things.

"Does the Phantom use a lot of power?" I ask. Normal pyrolysis might not give you much control but it is cheap; the biomass that becomes charcoal is the only fuel used. By contrast, this beast has large industrial power cables attached. I'm worried it might generate more carbon dioxide than it locks away.

"It does use power, and that's the price we pay for control. But we'll lock away twice as much carbon as we use. We're also looking into using the oil and gas produced by the pyrolization to power it," says Tim. "The idea is it becomes self-sufficient."

Later that afternoon Tim, Vicki and I take a trip to Picton, where Tim has his yacht *Faith* moored. It's another business venture as well as a personal pleasure—he hires the boat to groups of tourists for dolphin watching or simply to enjoy the majestic scenery of the north coast. The boat has some history too; it was once owned by Lord Shawcross, a British attorney general and chief prosecutor at the Nuremberg war trials that followed World War II. Shawcross successfully argued that "just following orders" wasn't an admissible defense when it came to collusion in torture, murder and genocide, paving the way for the new era of soldiers' individual accountability for their actions during wartime.

Tim takes us on a boat trip into Queen Charlotte Sound. As we leave the harbor, huge hills rise on either side of the water, magnificent and imposing against the deep blue sky. It's a glorious sight—nature's raw beauty painted on land and sea. The land to the left and right of us is covered with trees, and the sight of all this "biomass" inspires me to raise one of the objections to the proposed charcoal renaissance.

Some people worry that biochar might share a similar trajectory to crop-based biofuels, misappropriating land and taking it away from food production or disadvantaged peoples. In a largely out-of-character rant in the *Guardian* newspaper in 2009, the outstanding environmentalist George Monbiot characterized the idea as "woodchips with everything" and said it amounted to turning the planet's surface into charcoal. "[The] proposal boils down to this: we must destroy the biosphere in order to save it," he wrote. Monbiot is worried that when biochar activists talk of planting trees to turn into charcoal, they often refer to growing those trees on "degraded land" and that this is simply "the new code for natural habitat someone wants to destroy," land occupied by "poorly defended people whose rights and title can be disregarded." Advocates of biochar like James Lovelock, NASA scientist Jim Hansen, Chris Goodall and Australian global warming activist Professor Tim Flannery (who is on the board of Carbonscape) "should know better" wrote Monbiot.[13] (Just to put that in context, that's like someone telling Jimmy Page, John Lee Hooker *and* Andrés Segovia that they're not proper guitarists.) In fairness, Monbiot was a bit fed up with biochar hype and let rip. It's been heralded as a panacea, a complete answer to the global warming crisis—the cheap, easy solution. In reality, it's just one tool in the bag (although potentially a very powerful one).

Another concern is that as legislation comes into play insisting that more and more organizations offset their carbon emissions, they might turn to the cheapest solution: biochar, which, ironically, might lead to wide-spread deforestation. When eleven African countries asked the UN to consider international regulations to allow nations and companies to offset emissions using charcoal, there was outcry from some quarters. Rachel Smolker, a biologist and anti-biochar activist, gathered signatures from nearly a hundred and fifty concerned organizations to protest against the adoption of what she called a "charred earth policy." "It would require huge areas of land to be turned into plantations," warned Smolker.[14]

"Well, I don't think anyone's suggesting deforestation," says Vicki, slightly bemused. "It makes most sense to use agricultural and forest waste. I mean, there's so much of it, so many types around the world. Still, if the legislation is right, it could stimulate responsible reforestation."

"Planting trees in order to cook them?" I ask.

"Yes. That's the whole point. You plant fast-growing species, pyrolyze them and then replant. The amount of forest on the planet could end up being much higher than it is now. But Carbonscape's business is about maximizing what you can get out of *waste* biomass."

"What you can get out of waste biomass" turns out to be more than charcoal for soils, oil and wood gas. Tim talks about producing high-value chemicals that can be extracted and refined to provide further revenue for farmers. (He won't tell me which chemicals because the patents on the process are still being filed.) Vicki hopes that the control the Carbonscape process offers will soon yield "activated carbon." Activated carbon is an extremely porous form of charcoal that, because it has so many holes in it, presents a huge surface area to the outside world relative to its size. This gives it manifold uses from absorbing pollutants, to air purification, to water filtration and temporarily neutralizing the effects of poisons. You can even buy "flatulence filtering underwear" that passes your farts through an activated carbon filter to neutralize the whiff.[15] Yes. Really.

Half an hour after leaving the harbor, we drop anchor at Kumototo Bay, possibly one of the most tranquil places on the planet. It is startlingly quiet, the water barely moves. Only the occasional flapping of wings as a sea bird passes overhead disturbs the peace. After the whirlwind of new knowledge I've been stuffing into my brain on this journey, it's a welcome moment of calm. I can almost feel my brain slowing down a few clicks.

Vicki has supplied freshly cooked lobster that we eat with our fingers while sipping a locally made Chardonnay brought from belowdecks by Tim. I suddenly feel incredibly lucky to be here. Vicki and Tim are an ideal mix of easy company and sharp intellect. Conversation and ideas flow easily, like bubbles coming to the surface of a fizzy drink. Everything about this country and these people seems like a release.

Tim tells the story of an elderly German businessman and his wife who had hired the boat to go dolphin watching. It was one of those days where the creatures just weren't coming out to play, but the businessman was dogged in his determination to see them, requesting Tim try ever more unpromising spots when the usual viewing places had proved empty of our aquatic cousins. After about two hours, Tim anchored the boat for a drink and a snack in a quiet spot like the one we're now sitting in. "Do you want to continue looking?" he asked his guests. "Or would you just like to sit and be?"

"The funniest thing happened," says Tim. "He looked at me as if I'd suggested the most radical thing ever. 'Sit and *be*!' he said. 'Sit and *be*!'—as if it was the first time the idea had ever occurred to him. He called to his wife. 'Write this down!' he told her. 'Sit and be!' 'Do you want to keep looking for dolphins?' I asked, and he said, 'No! I want to sit and be!' so that's what we did."

We're all laughing, just sitting and being. The three hours I spend on the boat are some of the best of my entire trip.

That evening we have dinner in Le Café on the seafront at Picton, where I'm introduced to another codirector of Carbonscape (and Aquaflow), Nick Gerritsen, his gallery-owning partner Barb Speedy and a man called Don Binney who, I'm informed, is one of New Zealand's finest living artists, famous for his depictions of birds.

Nick Gerritsen is a firebrand, almost aggressive when I meet him. Clearly fiercely intelligent, he likes to make sure he's in worthy intellectual company. Physically he looks like a cross between a young Gene Hackman and John Lennon, circular spectacles sitting over eyes that give the impression of being searchlights seeking out opportunity. Or weakness. Nick is a "knowledge broker"—a man who puts people, ideas and money together. He's clearly not someone to suffer fools gladly. In fact, I get the impression he regards people who don't come up to his intellectual mark as walking versions of waste biomass. He's just come back from Australia, where he's reached the end of his tether with the nation's use of the word "mate" to refer to just about anyone. "Actually I'm not your bloody mate!" he rants. He's in a combative mood and wants to know whether my book is justifiable.

"What kind of an optimist are you?" he asks upon hearing the title.

"Yes, I hope you're not a *Panglossian* optimist," says Don Binney.

Bloody hell. After the relaxed conversation on the boat, it's suddenly got all gladiatorial. I've got a famous painter quoting Voltaire's *Candide* at me on one side and an *agent provocateur* fresh from Aussie-bating on the other.*

"Well, actually the book didn't start off with the word 'optimist' in the title," I say. "I just wanted to find out what was coming next. But when I told my agent Charlie about some of the stuff I was investigating he

---

*If you're not familiar with *Candide* (I wasn't), it's an eighteenth-century satire written by the philosopher Voltaire which lampoons, well, almost everything the author considered ill-thought-out or trivial. In it, the character Pangloss, who sees all history through rose-tinted spectacles, is used as a device to lampoon the optimistic philosophies of Leibniz.

said, 'this sounds great, this is optimistic stuff' and I realized he was right. There are people just getting on with incredible things that we don't really hear about enough. He put the 'optimist' in the title, not me. So, I'm a *convinced* optimist."

Don Binney seems happy with this. Nick considers for a moment. I decide to take a risk.

"You seem like you need to let off steam," I say. "Like you've got Argument Tourette's."

Vicki laughs uproariously and the mood immediately lightens. Nick lets go of his Australian frustrations and our conversation instantly becomes more like a good game of cards instead of the fencing match I'd feared. It seems no one in New Zealand takes themselves that seriously for long. I use the opportunity to ask Nick a few more questions about Carbonscape.

"Did you name the Phantom?"

"Yes. The engineering team wanted something more credible sounding but I figure people have long respected and loved the individual personality of machines."

"It's worked in our favor," says Tim. "It quickens bored journalists into interest. Let's face it, charcoal is pretty dull." True to form, the company's coming machines have equally comic-book names.

"The Phantom is really our research test bed," says Nick. "We're now commissioning a machine that can process biomass continuously."

"What's that called?" I ask

"The Rototron."

"That's actually based on the term 'magnetron,' the device that generates microwaves," says Tim. "So it makes sense."

"We're also working on a portable machine for rapid analysis of biomass so we know what it is possible to get out of it in terms of oil, gas, biochar and other products," continues Nick. "This will help quickly work out what you've got and select the cooking program to get the outcomes you want."

"I'm scared to ask . . ."

"It's called Zippy Bling," says Tim, smiling. "I named that! I was inspired by our bashed up Zip microwave. The name, like the microwave, has held out against the odds."

Despite the jollity and comic-book naming of machines, I'm wondering if I'm seeing the birth of a new industry here. Taking waste and turning it into useful products isn't a new idea (just ask The Wombles) but suddenly I'm seeing it writ large. With billions of tons of agricultural biomass essentially going to waste, there is an enormous opportunity. If Carbonscape or one of their competitors can help turn that into useful products and ways for farms and forests to earn extra income, we could see a rapid uptake of charcoal making, taking carbon out of the carbon cycle and realizing Lovelock's "one way we could save ourselves." Carbonscape wants to sell Zippy Blings and Rototrons to anyone who's got waste biomass.

"We want to get the technology as close to the end user as possible," says Nick. "The way I see it, this is a mining company. Except we're mining waste biomass off the surface of the Earth rather than ancient biomass from under it."

"Renewable mining?"

"Something like that, yeah."

Someone around the table suggests there should be a cocktail based on the title of the book, and much laughter ensues as we try to decide what should go in it and what it should be called. Nick calls over to Peter Schöni, the café's owner and barman, and we task him with creating "An Optimism."

"What do you want in it?" asks Peter, turning to me.

"Well, we've decided it needs tequila and champagne, but the rest is up to you," I respond.

Twenty minutes later, after some experiment, Peter returns with a champagne flute filled with a red drink.

"What's in it?" I ask.

"Take a sip," he says.

It's delicious; the perfect drink for a summer evening. I write the recipe down on a napkin which will end up framed in my flat: a shot of triple sec, a shot of tequila, lemon juice and strawberry pulp, shaken over ice and topped up with champagne. We collectively decide the drink shall be known as either "An Optimism" or "An Optimist" except here in Le Café in Picton, where it will have the special name of "A Schöni Optimist."

If you're ever there, please order one and tell Peter I sent you.

12

"A Little Bit of a Big Amount"

The nation that destroys its soil destroys itself.

—*Franklin D. Roosevelt*

t is the mid-1960s. Zimbabwe is in the grip of a civil war with its roots in white colonialism. As part of the war effort, a young scientist is put in charge of a "tracker combat unit." Even though he is hunting humans he is constantly observing and thinking about the condition of the land over which he pursues his quarry. He tracks people over game areas, tribal areas and commercial farms. Everywhere he inspects plants and soils, looking for the signs of human passage.[1] What he learns may one day help provide an answer to man-made climate change and improve the prosperity of nations.

It is late January 2009, and Michael and Anna Coughlan are about to surprise their new neighbors. Having just bought the farm Moombril near Holbrook, New South Wales, Australia, they proceed to sell off a large number of its assets including two houses, the shearers quarters, the shearing shed, the hay shed, the machinery shed, two garages (and the four vehicles in them), even some spare sheep dip. At the sale, whispers of incredulity can be heard passing among prospective local buyers. "What are they doing?" and "Why would they sell that? Won't they *need* it?"

Using part of the money raised, they purchase some fencing materials and two motorbikes. People think they are crazy.

It is today, and Tony Lovell is making a joke at my expense. My plane has just touched down at the tiny airport serving the farming community of Armidale, New South Wales. I shared the flight from Sydney with Tony's business partner, Bruce Ward, who is the more polite of the two.

Tony is a big chap, tall and solid. Yet, despite being a wide chunk of Australian manhood, he has something of the naughty child about him. The word that comes to mind is "cheeky." You get the impression that he's always on the brink of pulling a harmless prank. Bruce shares some of his friend's mischief but is more learned, more serious. In a way, Bruce is a more compact, more sensible version of Tony. Tony starts sentences with "Imagine if . . ." whereas Bruce will open with "The thing you need to realize is . . ." Tony will laugh out loud, Bruce will chuckle knowingly. But Tony's playful nature can't hide a deadly serious intent. He and Bruce are going to save the planet; he's just going to make sure they have a whole heap of fun doing it.

As Bruce and I load our bags into the back of Tony's LandCruiser, its driver tells me that our first port of call will be a farm that sacrifices Poms (Australian slang for Brits) in an ancient soil fertility rite. I am to be an offering to the gods. This sets the tone for the next few days. We will cover over a thousand miles as Tony and Bruce take me to a string of farms that are redefining the way we think about agriculture and that offer a model for mitigating global warming on an unprecedented scale, all while delivering sustainable increases to food production and bringing the cost of that food down. What's so striking about what they propose is that rather than using new technologies these farms are using an old one. "The thing you need to realize is it's a technology that's been around for millions of years," Bruce tells me. In the process I will learn what links a tracker in the Zimbabwean civil war to Michael and Anna Coughlan's asset sale in modern-day Holbrook.

Australian farming has become synonymous with drought. A decade of low rainfall, heat waves and wildfires has scorched much of the nation. Australians call it "The Big Dry" and it's brought acute hardship to rural communities. When the rains come, as they did to some parts of the country in early 2010, water runs over the parched surface, often resulting in devastating floods. It's as if the land is now so unfamiliar with water it no longer knows how to drink. The drought, it is said, remains deep in the soil. Here in New South Wales, Minister for Primary Industries and Rural Affairs Steve Whan remarked, "The recent rainfall events occur against a backdrop of decade-long rainfall deficits and record-high temperatures that have severely stressed water supplies."[2] Farmers across the continent are suffering. Agricultural debt has increased from just over ten billion dollars in 1994 to nearly sixty billion dollars in 2009.[3] Many of the farms we drive past on our journey are surviving on "drought assistance" payments handed out by the government.[4] Australia is worried about food production. Some believe the only way for the country to survive is to cut its population.[5] I'm here to investigate another solution.

"Imagine if you could take billions of tons of carbon dioxide from the atmosphere every year, safely, effectively, economically and immediately," says Tony as we get under way. "Imagine if you could do it in a way that also increases biodiversity, boosts food security, reverses the advance of the desert and improves rural communities."

I'd first heard this pitch in Manchester, England, before setting out on my journey. Tony had talked at an event called "the Manchester Report" held in the gothic splendor of Manchester Town Hall, a kind of grandiose (but deadly serious) game show for global warming–tackling initiatives. Over two days, twenty short-listed ideas had been publicly presented to an expert panel, and in front of a live audience. Tony's presentation stood out for two reasons. First, he's a very engaging speaker, combining authority with an approachable and light manner that makes you want to go to the pub with him. Second, what he said was extraordinary. The panel was "hugely impressed" and so was I.[6] It seemed too good to be true.

I introduced myself to Tony after the event. Bruce was there too, but couldn't get a word in edgeways as Tony and I instantly fell into the worst kind of joke one-upmanship, something we've slipped right back into here in Australia. He calls me "big fella" even though he's six-feet-four to my six-feet-one. Mocking his surname, Lovell, I call him "Love with two L's" (Love-ll). During long hours traveling, Bruce sits in the backseat alternately chuckling and dispensing wisdom but doesn't escape the barrage of puns and innuendos. When Bruce suggests a turning to Eubalong (pronouncing it "you-abbah-long"), for instance, Tony replies with "You have a long what? Try to keep it clean, Bruce." Everything they say about Australian humor is true. This goes on solidly for four days.

The jaw-dropping moment in Tony's Manchester presentation involved two photographs. "This is a typical ranch in Mexico," he explained, showing an image of an arid terra-cotta dustbowl with sparse vegetation and bare, compacted soil. Then he put up a second image showing a property awash with lush green vegetation, a verdant paradise. The contrast couldn't have been stronger, which made what Tony said next so astonishing. "This is the ranch next door. Same soil, same rainfall. These pictures were taken on the same day."

Tony and Bruce have a bunch of these photos, which they share with me over the coming days. Some of the most striking are aerial shots that show neighboring properties from above. The difference between vegetation and bare soil follows the line of the fences separating adjoining farms.

"What's the difference?" I ask.

"Management," says Bruce. "Just management."

About an hour after leaving the airport we arrive at Lana, the property of Tim and Karen Wright, where they farm merino sheep. Tim isn't convinced about man-made global warming. He thinks politics "has got in the road" and "there's a lot of hidden agendas going on at the IPCC, which means if you're not in there batting with them then bad luck." That hasn't stopped him and Karen from following Bruce and Tony's advice. We sit drinking lemonade on the veranda overlooking a green and fertile landscape. It looks more like Sussex than the images of the Australian bush I'm used to seeing on TV.

Tim and Karen ask me where I've been on my travels. When I mention my trip to Konarka, Karen immediately disappears inside. A moment later she brings out a sample of the company's Power Plastic, the same stuff I'd seen rolling off that "big printer" in New Bedford.

"We're a distributor for New South Wales," she says.

I guess in a country synonymous with drought it makes sound business sense for a farmer to diversify into selling solar panels.

"Is this a business for if the drought gets worse?" I ask.

There's an awkward silence. Clearly I haven't got it.

"We don't feel the drought like other people do," says Tim. "Our reservoirs are still three-quarters full."

The Big Dry isn't so big here it seems. But though Lana isn't feeling it, many of its neighbors are. Local farms that have been in families for generations are being sold off; it's just become too hard to make a living.

"Some are suicidal," says Tim grimly. He should know, he's worked as a counselor for farmers under stress.

Talk of rural suicide won't come as a surprise to anyone who lives in Australia. A widely quoted (but decade-old) statistic is that one farmer kills himself every four days.[7] Recent figures are hard to come by, but a 2009 report by the Australian Research Council concluded that the suicide rate was underreported and "all regional and all remote areas of Australia have alarmingly elevated national patterns of deaths as a result of acute external causes."[8] That's report-speak for "the drought is driving some farmers to kill themselves."

Tim tells me the rainfall has been low for *nine years*.[9] Yet despite this,

Lana is flourishing with high levels of livestock, tripling since 1980 from seven to twenty-two thousand "dry sheep equivalents," or DSE.

When I started this trip, I had no idea I would find myself getting to grips with sheep-based metrics. This feels a long way from my discussions with George Church and Eric Drexler about genetics and nanotechnology. DSE figures are a way to compare farm capacities and performance. "One DSE" is the amount of feed needed by a two-year-old merino sheep of the sort Tim and Karen farm. A pregnant ewe might be 1.5 DSE, a cow nursing a calf ten times that. (I'm not sure why the sheep got the monopoly on metrics.) Bruce tells me the usual labor ratio is one man per eight thousand DSE. Tim is running at one man to fifteen thousand.

"Tim and Karen are operating at full capacity with months of feed in front of them, where all around them is panic," says Bruce.

"The neighbors look at us with envy," says Karen. "They say 'you must have had a lot of rain here.'" She laughs, though without humor. "We've had the same rain as everyone else."

Something is clearly going very right with this farm, but it's not the only one—and Tony is eager we get on the road so I can see more. We've a long way to go.

We've been driving for about an hour when Bruce asks Tony to pull over.

"Let me show you something," he says.

As we get out of the car Bruce directs my attention to a paddock by the side of the road. The ground is almost totally bereft of vegetation— the sort of landscape you expect a man in a spacesuit to be walking across. I can't get over it. It's desolation.

"Is that the lack of rain?" I ask.

"Look down the fence," says Tony.

And then I see it, and from that moment on, I see it everywhere we go. Outside the fence there's grass. In fact, we're standing in it up to our knees. Inside the fence it's practically a desert.

"Things can grow just fine here," says Bruce.

Tony bends down and grabs a lump of vegetation. It comes away in his hand easily.

"This is dying though," he says. "That's the other half of the problem."

Back in the car Bruce tells me the story of a man called Allan Savory, who pioneered the farming methods employed by Tim and Karen and everyone else I'm going to see on this journey into Australia. Savory was also a tracker in the Zimbabwean civil war of the 1960s, but before that was the provincial game officer in the Northern and Luapula provinces of Zambia (then a protectorate of Northern Rhodesia).*

"Savory was trying to understand the continuing problem of land degradation, like what you've just seen in that paddock. His first instinct was that it must be something to do with the cattle, things that humans had inserted into the landscape in place of the natural wildlife. But it turned out it wasn't the animals, it was the way they were being managed."

Tony interjects, "If you go back in time, our grasslands were dominated by large herds of grazing animals, bison in America and wildebeest in Africa."

There are still a few places where you can see what Tony is talking about. The Serengeti, for instance, is one of the few remaining natural grasslands on the planet and is home to huge migrating herds of wildebeest and zebra. There is no beginning or end to their journey, but a constant clockwise pilgrimage in search of water and fresh grass. Every year roughly two million animals cover eighteen hundred miles. The herds stay closely packed as a defense against predators like lion, cheetah and hyena.

"What happens is the herd eats the grass but then moves on looking for the fresh stuff. In the Serengeti, that herd won't be back on the same ground for up to a year or so," says Tony.

---

*Savory fled to the U.S. following threats of imprisonment when he objected to the racist stance of the then white minority government, although today he spends nearly half his time in the country.

"That's important," says Bruce. "Savory realized there is a natural relationship between grasses and grazing animals. The growing buds are at the base of the plant and they need sunlight. If the plant gets too tall it starts to kill itself by hiding those buds in its own shade. It can't photosynthesize."

"That's why that grass came away in my hand," explains Tony Love-with-two-L's. "It's dead material. In nature the herd would have come along, eaten the tops off the plants, exposing the growth buds, and moved on. By the time they came back, the grass would have regrown. The animals have another go and so on and so on. The plant *needs* the animal to keep it alive."

"The problem with the way we farm livestock is we don't let them roam," says Bruce. "We split up big herds between separate paddocks and keep them there for way too long. With no predators they can wander where they like in that space. They don't stick together, which means they quickly eat all the grass in an area. Worse than that, the grass never gets a chance to grow back. An animal will have a go at it as soon as it starts sprouting. It's a vicious circle, with the animals destroying their own food. That's what you saw in that paddock back there: desert inside the fence and tall but unhealthy plants outside it—overgrazing and undergrazing side by side."

For the next four days I see this pattern of overgrazing and under-grazing repeated everywhere we go: miles upon miles of fences separate sparsely vegetated paddocks from verges of dried-out and withering foliage.

"Why doesn't anybody notice this?" I ask.

"Did you notice it?" responds Tony. It's a fair point.

Bruce, ever wise, says, "When you've grown up with something that's always looked one way, then you can't see it should look any different."

We spend the evening at the Overlander Motel in Gunnedah, the "Koala Capital of the World." Checking in I notice the establishment sells a product that only Australia could have come up with—the Stubby-glove. Dubbed "the greatest innovation in beer drinking technology since bubbles," it's a glove made of neoprene rubber with a built-in beer bottle holder.[10] ("Stubby" is Australian slang for beer bottle.) The idea, say the inventors, is that if you fall down, you won't drop your beer. I like the idea of a product that allows you to hold on to precious things during accidents, but only in Australia is falling over more of a trial if you've lost your *beer*, and not just your balance.*

"Is that for real?" I ask the receptionist, finding it hard to hide my amusement.

She ignores me and turns to Tony.

"What's his problem?"

"He's a Pom."

"Ah," she says nodding. Clearly this explains everything.

We eat in the motel restaurant, which serves up simple hearty meals heavy on meat. I use the opportunity to talk about the main reason I'm here: the climate change angle of Tony and Bruce's work.

"Look, even if there was no man-made global warming problem, what Tim and Karen are doing makes sense for their business. The good news is that the environmental benefits come screaming out the back end of it anyway," says Tony.

Those "environmental benefits" include the removal of carbon dioxide from the atmosphere at potentially massive levels as well as increasing biodiversity.

"There has been a huge decline in soil carbon levels across the grasslands of the world over the last hundred and fifty years which is directly related to the loss of vegetation. That paddock we stopped at? The soil carbon there would have been almost nonexistent."

Beers arrive and we elect to use our own hands to hold them despite the Stubbyglove option. As we drink, Bruce explains that grass plants

---

*The product had slightly higher ambitions when it was invented. The idea was born during a skiing trip to the mountains of New South Wales. Glen Krummel and his brother Leon watched their mate Jim, who had lost his arm in an accident years previously, struggle to keep his hand warm but his beer cold. The Stubbyglove was born, and as soon as the friends started using them, publicans began asking for branded versions. They've sold over a hundred thousand.

grow roughly the same amount of root matter as leaf matter—they're in balance above and below the ground. If the plant gets nibbled by a cow or sheep, it'll slough off a corresponding amount of root matter into the soil within minutes.

"Now a plant is fifty-eight percent carbon . . ." says Bruce.

". . . and carbon is good for the soil," I say, remembering what I'd learned on my trip to New Zealand.

"Very good," says Tony. "Clearly schooling has improved in the UK."

"But that carbon can't do anything to offset global warming, can it? I mean, nearly all of it is going to return to the atmosphere as the biomass rots."

"Right again," says Tony. "But the crucial words in your last sentence were '*nearly* all.' When organic material decays, it leaves behind a small amount of residue in the soil."

"Humus," says Bruce, giving that residue its scientific name.

"No thanks," says Tony. "I think I ordered the steak."

Bruce ignores this (I guess he's used to it). "That residue, the humus and charcoal components, stays in the soil in a natural system."

"Now it's only a small amount of the carbon that made up the plant," says Tony, "but a little bit of a bloody big amount soon adds up."

"Does this work in reverse?" I ask. "If there isn't vegetation, does the soil start to lose that stable carbon?"

"Yep. That's why conventional agriculture can bugger up a pristine ecosystem so quickly. That's what we've been driving through all day."

I'm arrested by a thought and look Tony straight in the eye.

"So let me get this right. You're saying that modern agriculture has released billions of tons of carbon into the atmosphere *from the soil?*"

"Yep."

"You're telling me that the culprit in global warming isn't just the Industrial Revolution, but the *agricultural* one?"

Tony looks over at Bruce. "He picks things up quick for an English fella, don't he?"

That's probably because I'm remembering something I read about the work of paleoclimatologist William Ruddiman, whose book *Plows, Plagues and Petroleum: How Humans Took Control of Climate* argues just this point. In fact, according to Ruddiman, agriculture is the reason we're not in an ice age right now.

Ruddiman's book reminds us that the planet is subject to a dance of astronomical cycles that affect how much of the sun's radiation reaches us. For the best part of the last million years, this means the Earth has experienced regular glacial cycles, with ice covering about a quarter of the planet's surface for a hundred thousand years at a time, interspersed with short ten-thousand-year "interglacial" warm periods. The last warm period should have ended two thousand years ago, says Ruddiman, but agriculture has warmed up the atmosphere and held it off. "Farming is not nature," he writes, "but rather the largest alteration of the Earth's surface that humans have yet achieved."[11]

According to the UN's "Livestock's Long Shadow" report, the total area of land given over to grazing amounts to twenty-six percent of the ice-free surface of the planet.[12] If, as Bruce and Tony are telling me, we've been managing that land in such a way that our soils have been losing carbon, then as agriculture has grown, so has atmospheric CO_2 as a result. This is not to underplay the crucial role of fossil fuels in accelerating that rise, but it is wrong to lay all the blame at their feet.

Some commentators worry that cattle burping methane (another potent greenhouse gas) into the sky is the biggest challenge to the climate from livestock and call for us to relinquish meat eating. Methane is twenty-three times more potent than CO_2 as a greenhouse gas and the UN estimates that livestock contribute eighty-six million tonnes of it to the atmosphere every year.[13] But if what Bruce and Tony are saying is right, getting rid of cattle could end up being a disaster. It's not that we have too many cattle, it's that we have too little grass. The soil needs the grass if it is to become richer in humus and therefore carbon. The grass needs animals if it's not to die from undergrazing. Working correctly together, the system is a huge natural "carbon pump" that can take massive quantities of CO_2 from the sky.

"The good news is that we only need to change the way the cattle move to make things better," says Bruce. "It's something we can do today and it more than pays for itself. The land is the biggest thing we can control. We can't manage the seas and we can't manage the air.* But we can actively manage the land. It's the single biggest asset we have in battling climate change and building biodiversity."

Tomorrow I'll find out just how big the impact could be.

After an early rise and a six-hour drive taking us over three hundred miles into ever more parched country we've come deep into an area known as the Western Division. Our journey had one stop—breakfast at the Jolly Cauli in Coonabarabran, a café that bizarrely doubles as an osteopath's. This mystifies me until I see Tony order his breakfast, which seems to contain the meat of an entire cow. "My grandma said to me, you don't get to this size by being fussy!" My guess is that the osteopath makes a good business dealing with back injuries caused by standing up too soon after eating here.

Now we've arrived at Etiwanda Station, a farm operated by Andrew and Megan Mosely in Cobar, one of the driest places in New South Wales. Soon after arriving Megan says, "When the TV wants a few shots to represent drought they film out here."

It hasn't rained in Cobar for five and half months but the Moselys are calm. Their farm is a beacon of prosperity in one of the most stressed parts of the country.

"The land hit its low point in the late nineties," says Andrew, handing out mugs of tea. "Back then we'd need fifteen acres to keep one sheep alive." In his eyes you can see the memory of bad times.

"The land got so degraded it was never going to recover under normal management," adds Megan.

"You were in a pretty dire state?" I ask.

---

*Bruce clearly hasn't heard about the work of Klaus Lackner.

"Yeah, we were not exactly sitting pretty. To be honest we were desperate. We needed to make some smart choices and in the end we had to completely change the way we managed the land."

"It was either that or give up and walk away," says Andrew

"Is it the same for your neighbors?"

"Many of them are living on government support. They're existing but they haven't got any kind of lifestyle. Their houses are falling down."

Megan and Andrew's house isn't falling down. In fact, it's a comfortable, well-furnished home with modern conveniences. Sprinklers keep the lawn outside a lush bright green, and healthy-looking pets lounge in the blistering heat, occasionally teased by the couple's two young daughters, Jessica and Emily. There's a small pig house near the property and I ask the girls the names of the two pigs who live there.

"Easter and Christmas," says Jessica.

"Those are nice names," I say, before realizing what they mean.

Andrew takes us on a tour of the property.

"We're seeing native perennial grasses returning. We're not sowing any grass seeds, they're just coming back naturally," he tells me.

"Where's the livestock?" I ask.

For a sheep farming operation I haven't actually seen any sheep. This strikes me as odd. By way of explanation Tony, who is doing his own bit to irrigate the land, says, "One of Allan Savory's central ideas is you get all your animals into big herds, like they would be in nature." Zipping up his fly, he continues, "What that means is that most of the property won't have animals on it for most of the time. They're all together in a small area."

"That's what gives the grass time to grow back," says Bruce.

"How do you keep them all together?" I ask.

"That's easy. You build lots of small paddocks. Leave them in each one for a day or two and then move them to the next."

"The biggest cost is fencing," says Andrew. "You have to put in a lot of fences to divide up the property that way."

After all my investigations into the cutting-edge of technology and medicine I'm struck forcibly by what Andrew has just said.

"One of the answers to global warming is *fencing*?" I ask. "This is the technology we're talking about here? You're saying pulling carbon dioxide out of the sky is a matter of fences?!"

"Yeah," says Tony, shrugging. "We reckon it's something even a Pom can get his head around."

Over dinner I turn the conversation to soil carbon levels. I want to know if the Moselys have seen levels rise the way Bruce and Tony are saying they should.

"We had some scientists up here from Trangie Research Station and the University of New England," Andrew tells me. "They reckon the carbon went up about fifty percent over three years."

"It was ridiculously low to start with, though," Megan interjects. "So fifty percent of very little is still very little."

Tony wants to know the exact figures. Megan pulls out a file and reads carefully.

"It went from 0.6 percent to 0.88 percent in the top fifteen centimeters of the soil," she says, her finger tapping the page. "Of course, it'd be higher now. We've had a bumper grass crop this year."

Lightning fast, Tony does a rough sum in his head.

"That's about five tonnes of carbon per hectare then," he says almost immediately.

That sounds like a lot of carbon for such a small percentage change but the figures stack up. An average soil weighs about 1.2 tonnes per cubic meter. A hectare (2.47 acres) is ten thousand square meters (107,639 square feet), so digging down a meter gives you twelve thousand tonnes of soil. An increase of just 0.28 percent in soil carbon as seen at Etiwanda Station therefore amounts to nearly thirty-four tonnes per hectare measured to a meter deep. Andrew and Megan's soil had only seen new humus penetrate the top fifteen centimeters. In time root matter will easily reach further, but for now it means that figure of thirty-four tonnes is reduced to a smidgen over five tonnes per hectare for the Moselys' farm. Tony was spot on. That's fast math.

In relation to climate change, though, there's another step in the calculation. Every extra atom of carbon that's now in the soil has got there thanks to photosynthesizing plants, which means it was originally part of a carbon dioxide molecule in the atmosphere. Because carbon accounts for only 27.3 percent of the weight of a carbon dioxide molecule, five tonnes of soil carbon is the equivalent of roughly *18.5* tonnes of CO_2 pulled out of the sky.

Extrapolating the increase in soil carbon over the Moselys' 26,000 hectares gives a figure of 480,000 tonnes of sequestered CO_2 from the atmosphere. To put this in context, this is equivalent to offsetting all the CO_2 emissions generated by over eighty-five hundred Australian citizens over the three-year period (and when it comes to per capita emissions, Australia is one of the world's worst offenders, with figures nearly triple that of your average European[14]). As Tony said in the restaurant last night, "A little bit of a bloody big amount soon adds up."

"For pretty much any soil type, if you increase organic matter by a full one percent to a depth of thirty centimeters, you sequester roughly a hundred tonnes of CO_2 per hectare," says Tony, taking a suck on a stubby. "It doesn't matter if it's degraded soils like here and you move from say 0.5 percent to 1.5 percent, or a Scottish moor where you've got fifteen percent organic matter already in the ground and you take it to sixteen percent."

Published research is full of examples showing how changes in grazing management can deliver increased levels of soil carbon. For instance, the Blackstone Ranch in Ranchos de Taos, New Mexico, sequestered over 13,600 tonnes of CO_2 over ninety-four hectares in just one year (that's nearly 145 tonnes of CO_2 per hectare per year), while doubling its livestock and increasing gross income by 395 percent.[15]

The UN's Food and Agriculture Organisation estimates there are 3.5 billion hectares of agricultural grasslands on our planet.[16] If we could increase organic matter in these by one percent, this would offset nearly twelve years of global CO_2 emissions. Some nations have already caught on. In July 2009, the Portuguese government introduced a multimillion-dollar soil carbon offset scheme based on dryland pasture improvement covering forty-two thousand hectares and compliant with the Kyoto Protocol. The scheme pays farmers to establish biodiverse pastures to increase soil carbon.[17]

Paying farmers for soil carbon increases could encourage more landowners to follow Andrew and Megan's example, and this is something Bruce and Tony have been tirelessly campaigning for.

After dinner the Moselys' eldest daughter, Emily, gives us a performance of "Mulga Bill's Bicycle," a bush poem by the legendary A. B. "Banjo" Paterson, which she delivers faultlessly. It's a nice family moment for a man who's a long way from home.

With the children in bed I ask a question that's been in the back of my mind since yesterday and one I'd nearly put to Tim and Karen.

"Could you not take over failing farms and do to them what you've done here? Are you tempted to buy other properties?"

Megan laughs. "If I had a bucket of money I'd buy the whole Western Division. You could do it; do up each farm, put in the fences, get the stock moving right. Yeah, you could take on places and restore them like you restore a house."

"If you had a bucket of money?"

"If we had a bucket of money."

Tony is smiling and I think I know why. There's a reason he can do complicated arithmetic in his head in double-quick time. You see, old Love-with-two-l's has another side to him. He isn't a farmer, or an agricultural scientist, or a soil carbon expert by profession. He's an accountant, and a very successful one.

A new day, another six hours on the road and a further three hundred miles bring us to the house of Ken Haylan, a retired businessman who now farms cattle at Hailiebrae outside the town of Blayney. Ken is in his sixties and manages the farm pretty much by himself. He's been following Bruce's advice for two years and the results should make any farmer worried about debt do a double-take.

On the journey down, Tony had told me another of his too-good-to-be-true statistics and I want to find out if he's pulling my leg. Over lunch I ask Ken about his figures.

"Tony tells me your costs of production are pretty low."

Ken considers for a moment.

"Yes. Although we're not fully stocked at the moment so it's a little higher than it could be."

"Can I ask you your per-kilogram figure?"

Ken tells me, my jaw drops and Tony smiles. You see, the average cost of production for a kilogram of beef in New South Wales is Aus$1.60.[18] Ken has just told me his costs are "about fifty-three or fifty-four cents."

He continues. "At full stock it'd be about forty cents."

"But that's a *quarter* of the average figure?! You're telling me you can sell your meat cheaper than your competitors can produce it and take a huge profit margin at the same time?"

"Something like that," he says nonchalantly. "Would you like another slice of quiche?"

I'm taking this in when Ken adds, "What unites all the farms you're seeing is that they are 'low input' operations. They mimic ancient grazing patterns within the parameters of a modern agricultural holding. It is the antithesis of most modern agricultural practice with high cost inputs and reliance on a chemical fix for a problem rather than seeing the productive base, the land, as a whole."

Since he's started following Bruce's advice, Ken says he has more grass in his paddocks when his cattle are leaving than other farms have waiting for their livestock entering a new field.

"Don't your neighbors wonder about that?"

"Oh yeah, but they'll never change."

"The problem is that farming's a traditional business and it stays in the family," Bruce explains. "Changing the way you farm is the same as saying that what your pappy and your grandpappy did was wrong, which isn't easy when they're still living on the property."

It's a common theme. Tim, Karen, Andrew and Megan had told similar stories. This is one of the reasons Tony wants Australia to follow Portugal's example and experiment with paying farmers for the carbon they sequester in the soil.

"It's a carrot instead of a stick," he says. "It's hard to change when everything's going to hell, when you've got to admit that the way you've been farming isn't working so good and all the while your dad's telling you to stick with tradition. But if you can pay them for the carbon, that gets them out of both holes—financial and emotional. The land improves, which is good for business, and they can tell pappy the change in management is to bring money in from something he never could have got paid for. It's not that the old man was wrong at all, it's just a new market: carbon."

"But even without carbon credits, they can see the difference just looking over the fence, surely?" I'm slightly incredulous.

"Well, these are special Australian fences," says Tony. "They're impervious to new ideas."

For once, he's not joking.

"Yep, you're never a prophet in your own dunghill," suggests Bruce sagely.

One day later, two hundred and fifty miles south, and I'm in an argument with perhaps the smartest farmer in Australia.

Graham Strong is not short of opinions or knowledge. He's vehemently opposed to genetic modification, for instance. I've mentioned the argument that selective breeding of plants and animals is a form of genetic modification and am now listening to an extremely educated rant about the difference between transgenic modification of species (the practice of taking a gene from one organism and deliberately inserting it into the genome of another) and selective breeding. "It's not the same and I hate it when people try to say they're equivalent!" Graham exclaims. I'd only raised it as a talking point, but our host has gone in for the kill. Tony is standing quietly to one side with the traces of a smirk on his lips. I think he's slightly enjoying the Pom getting an Outback-style bashing.

"Passionate" doesn't come close to describing Graham, who manages Arcadia at Boree Creek just outside the town of Narrandera. When he talks about the land, it's like he's referring to a brother, but don't let that give you the impression he's some tree-hugging hippie or the cliché of a city-hating farmhand. During his polemic about genetic modification, he reveals an impressive level of knowledge about many of the subjects I'd discussed at Harvard Medical School. Graham Strong is no country bumpkin.

When he talks, he has an urgent yet careful way of phrasing, giving him the authority of someone much older than his thirty-five years. He believes we need an urgent debate "about how we operate in society, about the relationship between our urban centers and how we produce our food, share water, share resources." Part of his contribution to that debate is getting involved in large public art projects that take an appreciation of the land to new audiences.

"You can't get upset with someone being ignorant about what you do if you've done nothing to tell them about it," he declares.

Some people, though, may not want to hear Graham's most controversial view: he doesn't believe in drought.

With the hardships being suffered throughout the Australian farming industry, this sort of talk can get you lynched. "I'm careful with the sensitivities, I understand them, but the word 'drought' means nothing without a cultural context," he tells me, his tone ever insistent.

"*Our* cultural context is the fact that our western lifestyle is at odds with the reality of our climate."

"But you can't deny the rainfall is low."

"Well, what's low? It's lower than we're used to, but there's still enough water if you manage the land right."

In a repeat of what I'd seen at Lana, Etiwanda Station and Hailiebrae, a tour of Arcadia earlier in the day had shown me a healthy farm with abundant crops of saltbush, a feed for Graham's sheep which delivers leaner, more succulent lamb. Part of the reason these farms are prospering, even during The Big Dry, is because when the rain does fall, their soils retain more water. According to the Soil Carbon Coalition, an Australian nonprofit that works to spread soil carbon knowledge "organic matter can hold four times its weight in water."[*,19] This means a one percent increase in organic matter relates to an extra hundred and fifty tonnes of water storage available per hectare. For a nation in the grip of The Big Dry, that makes a huge difference, and it's one of the reasons Tim and Karen were able to tell me "our reservoirs are still three-quarters full."

"The word 'drought' implies impotence," says Graham. "That nothing can be done. End of conversation. Well, no thanks."

One of the art projects Graham has been involved with is a film and music commission by UK-based artist Andy Ross. Andy spent a week at Arcadia a few months before my own visit. He says:

> I was asked to act as a voice through art for the issues of suffering and drought these farmers are subjected to day in, day out. I even wrote an e-mail to Graham before I went saying "I've heard you're having terrible trouble" which he felt completely patronised by. What I found was someone who simply wasn't going to succumb to the simplicity of the problem being framed as "drought." On the way home I kept thinking about the farmers I thought I was coming to see, the ones who are really suffering from drought, dealing with suicide, depression, family breakups and bankruptcy but I just hadn't come across anything like that during my time at Arcadia. I know the suffering continues to go on and it's still there, but what I found was a story about a farmer who was managing to adapt and was showing hope amongst all the despair.

---

* This fourfold estimate is conservative, they say.

I know what he means. I'm feeling a similar guilt. Everyone I'm meeting is happy.

"The thing is, you can adapt," says Graham. "It's not that you can't farm this land, it's just that you can't farm it the way you used to."

Our last stop is the farm Moombril near Holbrook, owned by Michael and Anna Coughlan, the couple who surprised their new neighbors by selling most of the property's assets as soon as they bought it. After another hundred miles and two hours of jokes, I'm finally looking at a cow. Having spent the last four days talking to livestock farmers and seeing only family pets and pigs named after the holidays they'll be eaten on, I've told my hosts I want to see some beef on the move, so Bruce, Tony, Michael Coughlan and I are in one paddock, looking into another which contains two thousand cattle—one of the large herds that mimic those that roam natural grasslands. These cows won't return to the paddock they're currently in for perhaps a hundred fifty days, giving the grass they're now happily munching on plenty of time to recover. In fact, the paddock we're looking at them from (and where they'll come next) has grass up to our knees. As far as I can see, there's greenery.

Michael would normally ride out here on one of the two motorbikes he and Anna bought with the proceeds of their asset sale. "It's just as easy to open a gate for two thousand cattle as two hundred," he says.

"Is it really as simple as it looks?" I ask.

"Well, in principle yes. The key to what we're doing is we've really dumbed it down, just simplified it and kept discipline with that. Because the animals are out here doing the work, the system kind of runs itself. We don't need to spend money and time on other inputs like fertilizers or pesticide sprays."

Ken Haylan, the man who'd blown my mind with his tiny production costs, had said something similar. "If you do it right, there's not much you should be spending money on," he'd told me. "You don't fertilize so you don't need a tractor; you don't need to spray because you don't have a major weed problem; and your cattle's health improves too."

Neither Ken nor Michael need to drench their cattle for worms. (Ken drenches cattle he buys once on purchase, "and that's only because I don't know where they've been.")

"It's tiring just watching some of my neighbors," says Michael. "I have been asked what I do all day."

"What do you do all day?"

"Well, we've just bought another farm to move over to this system, so that's keeping me pretty busy. But while I'm working there, putting in the fences and all the rest, I know my land here is going forward, which is a really nice feeling. It takes a lot of work to get the system right, get the fences right, the grazing plans sorted. It takes a good deal of thought."

He pauses.

"The thing is, in Australia and America we've absolutely pillaged our land. We've just fucked the whole thing. But I think we can turn it round really quickly."

We're on the long drive back to Sydney and, ironically, it's raining. I turn to Tony. There's something I've wanted to ask ever since our conversation over dinner with Andrew and Megan at Etiwanda Station.

"You're going to buy farms, aren't you?"

"Oh yes."

"How many?"

"We're starting with two and a half million acres. We're going to take large swathes of Australia and make it look right. Our minimum ambition is ten million acres. Then we'll move into other countries."

"Can you raise the sort of money you'll need to buy the land?"

"We are working on just that."

He's not lying. I meet Tony in London some months later, where he's holding meetings with large European pension funds who want to invest in long-term, sustainable assets. He's grinning like a Cheshire cat.

"Ah! Pomster!" he exclaims. "Let me buy you a beer."

Tony and Bruce want to save the planet. And they're going to make sure they have a whole heap of fun doing it.

13

"The Best Job in the World."

I know how men in exile feed on dreams of hope.

—*Aeschylus*

'm looking at one of the strangest things I've ever seen.

Eleven men are seated at three tables set in a horseshoe forma-tion. The top table is reserved for a single occupant, clearly the head honcho (there are two official flags to either side of him). In the middle of the arrangement is an impressive piece of decoration: a huge piece of coral. Solemn business is at hand. A document is passed from person to person and signed by each. The signatures are slow and purposeful, indicating perhaps the seriousness of the text they are attaching their names to. Or are people just more deliberate about signing things when their role is that of a government minister? Or maybe it simply takes longer to write when you're dressed head-to-toe in rubber. And twenty feet underwater.

It's been an odd day. Two hours ago, accompanied by a brace of jour-nalists and photographers, I'd stepped off the boat that brought me to Girifushi, one of the 1,190 islands nestled in the South Indian Ocean which make up the Republic of the Maldives. The sky, then as now, was a pure shimmering blue; the sand underfoot, soft and fine; the sea, a crystal-clear aquamarine teeming with life. A cooling but light breeze had rustled through the palm trees which sheltered us from the sweltering sky. If it hadn't been for the armed soldiers observing us with a mix of slight disdain and bemusement it could have been a scene from paradise. On the quayside, executives and technicians bustled around a tent full of expensive-looking audio-visual equipment with wires running out the back into the lagoon. In the water, scuba-clad cameramen bobbed about waiting for instructions while intense-looking *TV Maldives* staff rushed about with expressions that combined national pride with what I can only describe as the "I hope we don't cock this up" look.

The eyes of the world will soon descend.

Girifushi, like all the Maldivian islands, is less than six feet above sea level. So, if seas rise quickly due to global warming, there could soon be a lot less of the Maldives to visit. That said, the islands have a natural advantage over continental coasts when it comes to resisting the waves. Like many island nations the Maldives are coral atolls. (Indeed, the word "atoll" is derived from *atholhu* in the local language, Dhivehi.) Coral debris is constantly pushed onto the land by wind and water, and because the coral is always growing there's a continuous supply of island-building material. Paul Kench of the University of Auckland and Arthur Webb at the South Pacific Applied Geoscience Commission in Fiji looked at changes in the size and shape of twenty-seven atolls in the Pacific Ocean over a sixty-year period which saw sea levels rise, on average, about around two millimeters a year. What they found surprised many. The land area of most had either remained stable or grown, and just fourteen percent had lost landmass. "It has been thought that as the sea level goes up, islands will sit there and drown," says Kench. "But they won't. The sea level will go up and the island will start responding."[1] That's the good news.

The bad news is that sea level rise appears to have been accelerating since the onset of the nineteenth century, and no one's sure how fast the coral can keep up. Then there are concerns over the West Antarctic Ice

Sheet.[2] If that becomes unstable (as some climate scientists fear it could), sea levels could quickly increase by ten feet.[3] While we're at it, let's mention the melting of the Greenland ice sheet which has the potential to contribute another seven feet to sea levels if global warming continues unabated.[4] There's no way for the atolls (or the rest of us) to keep up with that. Another worry for the Maldives is that a sharp rise in sea *temperature* could kill that island-building coral, leaving the nation without its natural defense against modest rises. Just such an event happened in 1998 when a spike in water temperatures devastated the nation's reefs[5] (although there are now heartening signs of recovery[6]). The fragility of the republic in the face of global warming is obvious.

This is one reason why President of the Maldives Mohamed Nasheed (the man sitting in full scuba gear at the top table) has become one of the world's most vocal and respected voices on climate change-related matters, and it is he I've come to meet. Nasheed argues "If we cannot save three hundred and fifty thousand Maldivians from rising seas today, we cannot save the millions in New York, London, or Mumbai tomorrow."[7] To the world he says, "we are all Maldivians now" and compares his country to Poland in the Second World War: "a frontline state" in the battle against global warming.

But Nasheed isn't a pessimist, instead using the crisis to position the republic as a nation-sized laboratory of change, an example to the world of how we might arrest rising temperatures. "The Maldives is determined to break old habits," he told the United Nations. "From now on we will no longer be content to shout about the perils of climate change. Instead we believe our acute vulnerability provides us with the clarity of vision to understand how the problem may be solved."[8] Shortly after coming to power, he committed the nation to becoming carbon neutral within ten years. "We do this not because we can solve global warming on our own. We do this because we hope to lead by example. If the Maldives can become carbon neutral, bigger countries might follow. By doing the right thing and showing the way, we can make a far bigger impact than by blaming others for causing the problem," he says.[9] "We are determined to formulate a survival-kit, a carbon-neutral manual that enables others to replicate it, in order that all of us together might just about save ourselves from climate catastrophe."[10]

Too much of the debate over climate change has been debilitating, "about not doing things, about not emitting gas, about not going on holiday, about not having an ice cream," Nasheed told *Al Jazeera* in April 2009. "My feeling is this is the wrong way to go about it. We should be demanding we *do* things, do greener things, invest in renewable energy. Renewable energy is doable, it's feasible and will give you a handsome return."

Others don't see the point. Earlier today Nasheed's communication adviser, the impossibly youthful Paul Roberts, told me, "When the president announced the plan to go carbon neutral, we got a lot of pessimism in the international community, a lot of 'Why bother? You're going to drown anyway.'"

The president, however, is used to doing things against the odds.

Mohamed "Anni" Nasheed has led an extraordinary life, and he's only forty-two. In 2008, Nasheed's Maldivian Democratic Party (born during his exile in Sri Lanka and England) ended the thirty-year dictatorship of Maumoon Adbul Gayoom, a man who wasn't shy of conducting human rights abuses, particularly against those who didn't like his autocratic style of government. According to Amnesty International, detention without trial, torture and politically motivated assassinations were all part of the portfolio.[11] Recalling his experiences at the hands of his predecessor, Nasheed said, "I was imprisoned on sixteen different occasions and spent a total of six years in jail. Of these I spent eighteen months in solitary confinement. The thing that saddens me most about these experiences is that I was not able to witness the birth of my two daughters."[12] Stories of Nasheed's confinement are horrific. His hands and feet were shackled. His food was laced with laxatives and broken glass. He was often placed in a cell deliberately designed to heat up like an oven in the baking sunshine. He was regularly beaten.[13] Gayoom's men tortured him "to the brink of death" twice.[14]

This was a regime that hired London public relations firm Hill & Knowlton (reportedly paid £13,000 a month for their services) to present a tourist-friendly image.[15] But Gayoom wasn't helping. For instance, at a pro-democracy demonstration, he sent armed police to arrest a temporarily free Nasheed—who was subsequently charged with terrorism (apparently the loudspeaker he was using to address the crowd was "a weapon"). YouTube footage of the arrest soon made its way around the world (rather demonstrating the point I discussed with Vint Cerf about the Internet's role in exposing violence). One wonders what Hill & Knowlton had been doing for Gayoom. Surely in the PR for-hire rule book there's something that says, "Rule 46: When under the watchful eye of the international community and human rights organizations, try not to arrest popular democracy reformers. In particular don't send a pack of armed police into a large crowd (where lots of people have cameras) to drag away one unarmed man. It looks bad, trust us. Instead, we suggest a cocktail reception. P.S. Look out for that Internet thing. People love to chat over it."

The arrest, unsurprisingly, sparked mass public dissent, building on tensions that had been growing over the previous five years.* International observers voiced concerns that Nasheed would not receive a fair trial and, although it took another year in prison, he eventually walked free in exchange for a promise not to stir up revolution. But the revolution had already happened. Despite Gayoom trying to paint himself as a reformer (not an easy task when it's your own rule that needs reforming) Nasheed became the nation's first democratically elected leader.

On the occasion of Nasheed's election, Gayoom said, "My legacy is going to be introducing a modern liberal form of democracy: that is the

---

*For instance, in 2004, three thousand democracy protesters took to the streets of the island capital of Malé. (On Malé that's a *huge* demonstration.) Gayoom, true to form, sent in the riot police— hundreds were arrested, and the day became known as "Black Friday."

greatest legacy anyone can give"—which is a bit like Darth Vader saying, "I always intended to hand the Empire over to the rebels, I just got busy."

On taking office, the new president ordered the destruction of buildings previously used for detention and torture, and turned the multimillion-dollar presidential palace (an opulent disgrace for a nation where about one in five lives below the poverty line[16]) over to the judiciary to house a new supreme court. And yes, in true James Bond villain style, the former presidential abode *did* house a gold-plated toilet. If a bullion-encrusted commode doesn't cry out "I am a self-serving bastard," what does? And yet Nasheed allowed his former nemesis to walk free. "A test of our democracy will be how we treat Maumoon," he said.[17]

Nasheed and Gayoom may not share an interest in gaudy bathroom furniture, but they do both understand the importance of PR, although Nasheed's a damn sight better at it than his predecessor. As an ex-journalist (it was criticism of Gayoom in the magazine *Sangu* that first brought him into direct conflict with the previous regime), he has used his media savvy to bring the plight of the Maldives to the international stage. His boldest PR stunt to date is the event I'm now looking at through my snorkel mask: the world's first underwater cabinet meeting.

I'm here because the Maldives encapsulates the climate change issue. What happens in this island republic will be instructive to us all. It will be one of the first nations to cease to exist if sea levels rise too quickly. (Nasheed has already talked about setting up a "sovereign fund" to buy land elsewhere should worse come to the worst.) At the same time, the country is leading the way on creating a carbon neutral economy (no small feat when the nation is made up of over a thousand tiny islands, nearly all of which get their energy from diesel generators). Then there's the Maldivian economy, dependent on industries built on a model of conspicuous consumption. The nation's carbon footprint is dominated by tourism and associated aviation. The country also demonstrates the carbon divide between "developed" and "developing" nations. Resorts renowned for their opulence host small numbers of wealthy tourists who consume luxury foods flown in to meet their demands. Sumptuous villas are filled with every electronic convenience contributing to carbon footprints that dwarf those of the native populace. By contrast, neighboring nonresort islands often have populations of thousands, many of whom live in poverty. "We have a situation in the Maldives where you have a very poor third world island which is next to a very rich European island," Nasheed has said—a legacy of the Gayoom era which saw too many tourist dollars going on gilded toilets and suchlike in preference to social care. One of Nasheed's initiatives, therefore, is encouraging resorts to directly link their economies with neighboring islands by buying labor and supplies.

Another reason for coming to see the president is it allows me to take a step back from the science and consider, for want of a better phrase, "the real world." I've seen a good deal of techno-optimism, much of it well founded, but I want to talk to someone who understands more than most the emotional and political practicalities of change and how hard it can be. How, I wonder, can we turn knowing new things into doing the *right* thing? We may have solutions to our problems, but that doesn't mean we always use them. To paraphrase one of Winston Churchill's alleged sayings, "Humanity can always be trusted to do the right thing . . . once all other possibilities have been exhausted." I'm guessing a man who's gone from prisoner to president might have a few ideas on how to get us on the right path before exhausting all other options.

To save his nation Nasheed needs the rest of the world to make changes to the ways they generate and consume energy. In an impassioned speech to the UN assembly, he was unequivocal: "It's crystal clear to us. . . . If things go 'business as usual' we will not live. We will die. Our country will not exist."[18] In the same way he proposes linking the economies of resorts and neighboring islands, he seeks to unite those who emit the most carbon with those who might be the first to suffer the consequences.

Thanks to Paul Roberts, I'm lucky enough to be one of just four people with no official capacity allowed into the (deliciously warm) water. I swim around the perimeter of the meeting careful to avoid numerous scuba-clad cameramen and wires trailing into the lagoon from the shore.

Cabinet ministers pass the document to each other in an orderly procession of aquatic cordiality, and one by one they use their waterproof markers to sign it: a call for nations around the world to cut greenhouse gas emissions that will be delivered to the forthcoming Copenhagen climate change conference. The odd familiarity of the exercise, I realize, is the thing that will make it great publicity. It is both ordinary and extraordinary simultaneously. Who *isn't* intrigued by a government meeting taking place underwater? I also warrant it's the only time you'll see an entire meeting dressed in tight-fitting black neoprene that's not something to do with the English aristocracy.

The fish seem largely unimpressed, as if having a bunch of cabinet ministers, a president, a brace of support divers, underwater camerapeople and some office furniture in the water is an everyday occurrence. Because no one can talk, the whole affair is carefully choreographed using hand signals, with each participant following a waterproof "order of service." "President signs statement. Cabinet pass the slate one by one. President signals cabinet ascent" . . . and so on.

Everyone sticks to the plan (I'm guessing a late amendment would be tricky), and given that most of them have had to take diving lessons in preparation I suspect they're just happy to get through the whole thing without drowning or making a tit of themselves.

All in all the meeting lasts about twenty minutes. The ministers slowly rise from their watery seats and make their way back to the lagoon's edge. I find myself swimming next to the president. His head turns my way and I must look startled because he makes the underwater hand signal for "Are you okay?" I signal back to assure him I am. I'm more than okay, but there isn't a hand signal for "Bloody hell! I'm at an underwater cabinet meeting in the Maldives! *How cool is that?!*"

Reaching the shore, myriad camera lenses and microphones await us. The world's press are clamoring for the best vantage point at the water's edge, launching into a barrage of questions that Nasheed answers from the water, being careful to link the threat he sees to the Maldives with that faced by the rest of the world. There's also some good-natured banter about the benefit of having a cabinet meeting where none of your ministers can talk. Asked whether underwater meetings might become a regular feature of the Nasheed administration he replies, "The whole idea is that this *doesn't* become a regular feature."

The next day I walk the perimeter of the capital island of Malé, a view of the Maldives few visitors ever see (most transfer directly from the airport to their chosen resorts) and it's one that is revealing.

By day, Malé's two square kilometers host somewhere between 90,000 and 150,000 people (definitive figures are hard to come by). Even by the lowest estimates this makes it one of the most densely occupied cities on the planet. In stark contrast to the spacious whole-island resorts Malé is a warren of tight noisy streets, filled with the buzzing of thousands of motor scooters all owned by people who drive like Konarka's Tracy Wemett. On acid. Buildings are crammed together, and it's hot, hot, hot. Looking at this mass of humanity crammed onto such a small island, my thoughts turn to the issue of population growth and its effect on the planet's resources. Here in the Maldives the population has increased sixfold in the last century.[19]

In the year 1 CE, the population of the planet has been estimated at just 231 million.[20] By 1804, there were one billion of us. But then we really got going. It took us just one hundred and twenty-three years to add another billion to the total. The next additional billion arrived in just thirty-three years. Fourteen years after (in 1960), we'd added another billion (to bring us to four). The next billions came quicker still, taking thirteen and twelve years respectively. By the turn of the millennium, we'd passed six billion. If we carry on like this, accepted wisdom is that the population of the planet will soon outstrip its resources.

By the docks I find an example of mankind's consumption. Hundreds of boats offload strong-smelling catches of huge yellowfin tuna. The fish are dumped unceremoniously into pots with their heads down (and thus obscured), their huge tails sticking into the air. The result looks like a strange cross between a cactus and a hat stand. In a tiled hall on the waterfront I witness the fish being gutted and prepared for sale. Huge knives expertly gouge out lidless eyes, heads are ripped off, spines are removed in single, swift and well-practiced movements by men smoking cigarettes and made indifferent to their butchery through repetition. I guess ripping the head off a tuna becomes as mundane as processing an invoice if you've done it enough times. Around me there is a huge amount of trade going on.

I ask one of the fishmongers if he was born in Malé and he says, "No, but we move here because if you don't work on a resort, this is where the jobs are."

Moving to the city to find work is the other key trend in population. Back when there was one billion of us, just three percent lived in an urban environment. Today over half of us do. The UN's *World Urbanization Prospects* report estimates that the number of us living in towns and cities will nearly double by 2050. "Virtually all of the world's population growth will be absorbed by the urban areas of the less developed regions," it says.[21]

In 1798, foreseeing a catastrophe where human numbers would outstrip the carrying capacity of the planet, the Reverend Thomas Robert Malthus famously wrote (and over nearly thirty years continually rewrote) his influential "An Essay on the Principle of Population." In it he predicted that food production would eventually fail to keep up with the population. "It is an obvious truth, which has been taken notice of by many writers, that population must always be kept down to the level of the means of subsistence," he concluded.[22]

Malthus was both right and wrong. He was right because food production must, of course, keep pace with population if we are not to starve. He was wrong because so far, as the population has grown, we've managed to find ways to increase agricultural yields to match our needs. That is not to say that in some areas of the world there aren't shocking levels of malnutrition. This, however, is a problem with distribution, not production. Despite the dramatic increase in human population since the 1950s, world food production per capita has increased. That's a stunning achievement. There's more food per person than there's ever been. Our crime has been that we have not distributed either it, or the means of production equitably. Even still, according to the UN, "There is a global food security situation that is steadily improving, with a consistently increasing global level of food consumption per capita."[23]

The big worry is that the effects of global warming will greatly reduce our capacity to grow food as crops wither under the onslaught of rising temperatures, which is why what I saw during my week in Australia with Bruce and Tony could turn out to be staggeringly important. Ironically, one way to continue to increase food production in a sustainable way could be to return to ancient practices. But can we continue to raise food production ad infinitum as our population continues to grow? The good news is we won't have to.

The statistician Hans Rosling from Sweden's Karolinska Institute tells the story of how many of his students wonder if keeping the poor alive is really such a good idea.

"In a time when we know the pressures on the environment are growing, my students tell me 'population growth destroys the environment so poor children may as well die.' This is a statement they don't make in class, but afterwards. They say, 'Why save the lives of all these small children? Because if they survive, we'll be even more people, we'll destroy the environment and they will emit carbon dioxide and it will be even worse and then we will all die in the end.'"

The problem with this logic, says Rosling, "is not that it's not moral, it's that it's wrong"[24]—and here's why.

As the population is growing it seems to be stabilizing. The rate of growth is slowing down dramatically and looks set to stop (and possibly reverse) sometime around 2050. Already the population in many countries is falling. The important figure to keep in mind is 2.1. This is the number of children, on average, every woman needs to have to keep the population static. (The 0.1 accounts for the fact that some of us never have children, either because we choose not to, suffer from infertility or die before we get chance.) Between 1950 and 1955, the number of countries where the fertility rate was less than the crucial 2.1 figure was just five. In the period from 1995 to 2000, that number had risen nearly twelvefold to fifty-nine.[25] According to the UN, countries who weren't replacing their population in 2010 included the United States, Canada, Russia, China, Australia, New Zealand, every country in Europe, Chile, Brazil, Sri Lanka, Thailand, North and South Korea, Tunisia, Japan and Singapore. The Maldives isn't replacing its population either, with a total fertility rate of 2.[26]

Countries with the highest number of births and therefore rapidly growing populations are usually the poorest, with Africa dominating the table. Across Africa fertility rates of between 4 and 7 are common, but crucially these too are dropping. Taken worldwide the human race's fertility rate has fallen from 4.92 in the 1950s to 2.56 at the beginning of this century. By 2050, the UN's "median variant estimate" (a sort of "best guess") is that the figure will dip below the all-important 2.1 replacement level across the entire world population.[27] This means we'll stabilize at just over nine billion people, and then possibly fall in number. But why is this happening?

The contributing factors seem to be falling rates of child mortality, increased life expectancy, more urbanization, improved access to birth control and the growing emancipation of women, all of which are in some way linked to increased prosperity. The first two reasons seem counterintuitive. If both us and our children are living longer, surely that would increase population, giving us more time to breed and creating more offspring to do the same? The figures, however, show the opposite trend.

One theory is that when you know fewer of your children are likely to die, and those that survive will live longer, you don't feel compelled to have as many. Second, prosperity is linked to increased choice, not only in whether to have children or not (through access to birth control) but also in what to do with your time. With the health of our children assured we can choose to spend more of our time being entertained and educated, and dedicating more time to educating and entertaining them too. With a few notable exceptions, prosperous nations seem to automatically reduce their birth rate (as the list of countries above amply demonstrates).

The rise of cities also has a role to play. One of the results of prosperity is urbanization and this in turn also seems to encourage us to breed less. According to George Martine, lead author of the UN Population Fund's *State of World Population 2007* report, eighty to ninety percent of economic growth takes place in cities, which drives prosperity, which reduces the number of children we have. Martine also makes the point that the increased density of city populations "makes it easier to provide social services, or services of any kind: education, health, sanitation, water, electrical power. Everything is so much easier, much cheaper on a per capita basis." The report also found that female emancipation is more likely to occur in the city, again reducing birth rates.[28] The city is where many go to make a career, often delaying starting a family. Property is also more expensive, meaning that even though you may be wealthier than many of the rural population, you have a lot less space in which to put your brood. And, of course, in the city the distractions of entertainment and education are ever greater. All this means that across the world, fertility rates are lower in cities than they are in rural areas—which, as the world becomes ever more urban, confers yet another downward pressure on the birth rate. Mankind is like a young boy in a coming-of-age movie. As soon as it got big enough, it went in search of the bright lights to see if life might be better and to make its fortune, rather than staying at home, marrying a local girl and having three kids.

Beyond all this, cities are also good for biodiversity because they attract subsistence farmers away from the land. Subsistence farming might have a romantic image of honest rural folk in tune with their environment but the clue is in the name. Subsistence, by definition, is "The action or fact of maintaining or supporting oneself at a minimum level." Subsistence farming is characterized by extreme poverty, along with overgrazing and tilling of the land that has seen soil carbon levels plummet along with the reduction of biodiversity as people use the natural ecosystem for resources, a key example being the use of wood as a fuel for cooking and heating. When humans leave, natural ecosystems return.

The UN's *State of the World's Forests* report shows that as countries prosper and urbanize, so their forests regrow. In Europe, forest resources "are expected to continue to expand in view of declining land dependence, increasing income, concern for protection of the environment and well-developed policy and institutional frameworks." In Asia and the Pacific, "forest area will stabilize and increase in most of the developed countries." In parts of Latin America "where population densities are high, increasing urbanization will cause a shift away from agriculture, forest clearance will decline and some cleared areas will revert to forest." In Africa, however, deforestation continues apace.[29] Overall though, as we move to the city we leave behind ecosystems that regrow. Some estimates suggest that for every acre of rainforest being cut down, over fifty acres of secondary forest are regrowing.[30] That's pretty incredible, don't you think? (It is, however, important to stress that regrown "secondary" forest in no way matches up to the biodiversity lost when ancient rain forests are destroyed.)

This teeming metropolis I'm walking through is therefore an engine of renewal. Hans Rosling's students are dead wrong. Those "poor children" must live, move to the city, and prosper and in doing so they will, in just a few generations, stabilize the world populace while allowing many ecosystems to flourish.

It's a nice thought to have as the sun sets on the Indian Ocean.

Close up, the first thing you notice about Nasheed are his eyes. Despite being a dark brown, they seem infused with light, and I get the feeling that behind them and at every moment a great deal of work is going on, as if they're small windows that overlook a huge intellectual conveyor belt. He wears a smart but not flashy suit and, despite the heat, a tie. The impression is one of care rather than fastidiousness. There's also a sense of nimble grace about him. Perhaps most surprising is his candor: there is little of the guarded talk typical of career politicians. By contrast, he seems rather fresh and disarming. Early in our conversation he says, "You know I'm always told 'be cautious—not to do that, not to say this.' I end up saying something 'wrong' every week and they don't like it." He smiles. "But I have to go on saying what I believe in."

The "they" in question is the governmental machine he's inherited, which clearly frustrates him. How does he cope? I ask. He laughs and says something I really don't expect.

"I think what's helping me is Tom Sharpe"—an out-of-the-blue reference to the English satirical author whose comic novels are famous for graphically and lewdly lampooning authoritarianism. "The comedy of it all—of government and endless meetings. You'd be amazed at the kind of 'work' I do," chuckles the president.

The Copenhagen climate change conference (which he attends two months later) must have been a special kind of hell for Nasheed, yet he was hailed as "the real hero" of the conference by Danish prime minister Lars Rasmussen. "The Copenhagen Accord is a long way from perfect. But it is a step in the right direction toward curbing climate change," said Nasheed, before returning home to get on with leading by example.[31] I get the strong impression that here is a man who understands better than most how to negotiate the seemingly impossible, and who has the patience to endure and guide an almost infinite progression of baby steps to a resolution.

"When you reach a dead end in trying to convince someone or trying to do something, it would be best to give it a moment, and give it some thought. There's no value in just banging on. There's always more than one avenue to any destination. Even in a dead end, when things get really, really bad you have to keep going—however badly you suffer, whatever losses you incur, you just have to keep going. You have to make a tiny step. Don't just be stuck with a single option."

By example Nasheed recalls his interrogations at the hands of the last regime's chief of police, Adam Zahir, which became a battle of wills. "I can see many would say there is no point reasoning with him, but there is nothing else better than reason," he says. "And despite his reputation, you could see he was wobbly."

Zahir was the only person Nasheed asked to resign when power was transferred. I'm fascinated by how he works with those who remain in the military and the police force that formerly worked to oppress him. He answers with a typical optimistic pragmatism.

"I think they, being very hard people, with their military background and minds, actually look up to me because I did not capitulate." Suddenly he becomes stern, and the pencil he is holding becomes a pointer. "They tried *their best*," he says stabbing it forward. "They tried to get me to capitulate." A short silence hangs in the air. I've read about what they did to him and I don't know what to say. He lightens. "You know some of the police officers I work with now were my personal interrogators." He's almost amused by the irony, but then a hint of metal creeps back into his voice. "They know who they're talking to."

"Are there parallels between a man in solitary confinement and a tiny nation in the midst of the world's biggest problem?" I ask. "Do you think something in your solitary confinement prepared you for this role?"

He looks at me squarely and I worry for a second my question might be taken as making light of his ordeal, that I'm trying to spin it as a useful experience and have therefore trivialized what happened to him. But instead he exclaims, "I think you're very right there! Yes! If you can muster the faculties to survive in solitary for long periods of time, you must have some mechanisms, some tools upon which you can build a strategy for stopping global warming. Very true." He points, not at me, but as if at some philosophical target hovering between us. "I was just one person right in the middle of some huge, very sophisticated machinery. And we as the Maldives are in solitary, surrounded by bigger nations and big countries with huge achievements and we are just probably nothing, but still . . ." He shrugs. "We have our ideas. We want to survive. We are not asking for much."

From this point on it becomes hard for me to separate the man from the nation. When he speaks "I" and "we" become interchangeable. When he recalls the struggle for democracy or talks of the current battle against global warming, his language is also transposed freely between the two (indeed, he often slips between time frames when answering a question). For him, it seems the two are not separated, but instead linked events in a continuous mission to help his homeland flourish. Perhaps this is why so many people find Nasheed a compelling negotiator—an ability to come across not just as a representative of his country, but an embodiment of it. In our time together I certainly begin to get this feeling. It's as much in the way he rebuffs hard-line Islamists (who have publicly criticized him this week for removing his wetsuit and exposing his chest at the close of the underwater cabinet meeting) as it is his views of political and envi- ronmental realities. "No sane Maldivian would think you could be in the water with anything on you. What are they talking about? Have we ever gone swimming with a T-shirt on? No! So why should the president? That is not the Maldives." The message is something like "I'm a Maldiv- ian first, and the president second."

Climate change is, he says, "the biggest challenge we will ever face. Not terrorism, not piracy, not drug dealing; nothing compared to this. So we really need to try and do something about it. No matter how small or however insignificant we may be."

Given the severity of the threat he sees to his nation, how does he remain optimistic?

"With the belief that there is hope, that there is a bright future. By seeing a picture other than the very fearful picture that is staring you right in the face." He slips time frames and we're back in the democracy battle. "What I would try to do is imagine another country, another homeland, another time, other circumstances."

"It's all about having the right vision?"

"I've always been optimistic. I feel if you can show the light at the end of the tunnel it's bearable to move out from difficult times and situations. I know this is huge odds, but if you look at the situation the Maldives were in five years ago, most people would have said, 'What is the point with the democracy movement? We are wasting our time.'"

The fact I'm having this conversation with a head of state who was a former political prisoner, while sitting in the presidential office, rather amplifies the essence of his argument.

The big picture, the goal, the brighter future, optimism that things can be achieved. All admirable ideals but how, I ask him, does one hold on to these when you're embroiled in the "endless meetings" and the "comedy" of government he talked about earlier? By way of an answer, he sighs and puts his hand to his temple in a gesture of comic resignation. This inspires me to ask if, in a strange way, he misses his days spent under house arrest where, if nothing else, he had plenty of time to think. He smiles.

"I really do," he says almost wistfully. "I'm surviving from the reserve— and my feeling is one might only be able to survive for five years on that— and this is probably why five years is a natural term for anyone to be a leader."

"Your brain is too full with the day-to-day now?"

He leans forward. "You don't get the bigger picture," he says urgently. "You lose the concepts. So then you get hold of processes, frameworks, strategic plans, matrices—it's all very good . . ." He trails off. "But if you don't see the bigger picture . . ."

This is the essence of leadership. Keeping to a simple vision, even when things get complicated. I'm reminded of something Bruce Ward said to me in the Australian outback: "Keeping things simple is never easy."

The president points to the clock indicating that our time is nearly up and I decide to use the minute or two I have left to do a bit of advocacy for Klaus Lackner and his carbon scrubbing technology. As I explain the potential of Klaus's work I can see his brain working. His eyes go up and to the right. He leans forward. How much money does Klaus need?

(Nasheed tells me he is hoping to put aside a hundred million dollars each year for investment in carbon neutral projects). I repeat the figure Klaus gave me of twenty million dollars to take his technology to the next stage—a design that can be rolled out worldwide. Suddenly I find myself suggesting to the president that maybe one of the deserted islands in the Maldives might act as a good demonstrator for the technology, to show the world its potential. As an act of cheek it goes beyond anything I've ever done. The bloke from New Cross has just overstepped the mark, suggesting what a president should do with a part of his nation. Except the president says, "So he should come down here, and we could give him some room . . ."

Should I put him in touch with Klaus?

"Please do!"

And what would be the best way of doing that I wonder?

"I think the best way would be to get in touch by personal e-mail," he says, scribbling his contact details on a piece of paper and handing it to me. I'm dumbfounded. A president has just handed me his private e-mail address. (It's not a government e-mail account, it's hosted by one of the world's popular web-mail providers.) I want to phone my mum and shout, "Mum! Guess what?!"

I have one last question. What's his one tip for approaching the future?

"Never give up hope, you know? Never give up. Just keep moving." He pauses. "Tomorrow must be better. Tomorrow is better."

Our interview is over and we pose for a photo together as I thank him for his time. He smiles readily and, ever the diplomat, thanks me for my questions. And then he is gone.

Back on the humid streets of the capital, I have a huge grin on my face. Nasheed is a lightning rod for optimism, and it's hard not to feel better about the future after spending time with him. And then it strikes me forcibly that it's not just him, but all the people I've been meeting on my journey. All of them are inspired by what can be done, all of them are doing something to make it real, all are motivated by a desire to improve the human condition and none of them is waiting for permission.

During our talk the president had said, "Thoughts are real. Once you have given thought to something, it then becomes material very often and quickly."

He is, of course, right. A few weeks after I leave the Maldives, Nasheed announces the government has commissioned an offshore wind farm comprising thirty large turbines delivering power via a network of underwater cables. When it's finished, the plant will provide forty percent of the nation's electricity and reduce its carbon emissions by a quarter.[32]

You get the feeling he's only just getting started.

I'm not used to luxury. Or rather, I'm not used to the luxury of the rich. (I understand the luxury of a really good Yorkshire pudding as much as the next man.) And yet here I am at the Soneva Fushi resort on Kunfunadhoo Island. Regularly voted one of the best resorts in the world, it's the *crème de la crème* of holidays for the rich and famous who want to get away from it all and enjoy a quiet, understated opulence. I could never afford to stay here, but am getting a glimpse of how the other half live as a guest of the owners Sonu and Eva Shivdasani. Arriving on the island, I'm assigned my own personal "Man Friday," the impossibly smiley Hannan, who tells me that the maximum occupancy of the resort is about a hundred and fifty guests who sleep in private villas (each with its own swimming pool), attended to by a staff of nearly four hundred. The smaller neighboring island of Maahlos, know locally as "Mosquito Island" and where much of his family live, has a population of three thousand.

On one level, Soneva Fushi could be seen as everything that's wrong with the planet—rich westerners with individual carbon footprints the size of Donald Trump's ego holidaying next to unseen poverty. But the hotel chain that owns the resort has committed itself to be carbon *negative* by 2020, and that includes offsetting the flights of all its guests via a scheme to finance community wind turbines in southern India.

The tall and willowy Anke Hofmeister, Soneva Fushi's resident marine biologist and environmental manager, takes me on a backstage tour.

"This is the best job in the world isn't it?" I ask Anke.

"Yeah. Except there are no single men. Other than that it really is," she replies, smiling.

Anke shows me how waste packaging and food is mixed with the island's sandy soils, increasing their carbon content and allowing the island to grow most of the vegetables used in the kitchens and cutting down on air miles while providing super-fresh produce. The remaining food waste decomposes in a "biogas plant" producing fuel for cooking and power generation. Nonfood waste is turned into biochar which is used to improve soil fertility further. Waste glass is crushed to make building material. Anke shows me how excess heat from diesel generators is used to warm water for guests. A small solar farm being installed during my stay will soon provide ten percent of the island's energy needs, cutting down on that diesel. It's all pretty impressive, almost a summary of the solutions to global warming I've spent the last weeks investigating. "Of course," I reason, "it's easier to be a responsible corporate citizen when your clients are the super-rich." There again, if every resort in the nation followed this example, the republic would already be close to achieving Nasheed's objective and I'm suddenly struck by the ugliness of my own cynicism. I wander off to look at the sea and have a think.

Sitting on the beach I'm overcome by a feeling of complete exhaustion. My journey has been mind-blowing and my brain is ready to fall over. While I was in Boston, I'd caught a one-woman show by Gioia De Cari about her time as a student at MIT. "Getting an education from MIT is like taking a drink from a fire hose," she'd said. That's how I'm feeling about my trip. I expected the future to be different, but I didn't expect it to be so radically different. Everything is colliding inside my head. I'm disoriented in paradise. I've seen revolutions in medicine, biology, robotics, nanotechnology, energy production and communications that could change everything, if we can avoid a climate catastrophe (and I've seen solutions to that too). My brain feels like an old pocketwatch that's been overwound and is about to explode cartoon-style, scattering springs and cogs onto the hot sand.

Like Nasheed fighting not to be overwhelmed by the government machine he's inherited, I'm overwhelmed with new knowledge. How does all this fit together? Where are we going? How do I find "the big picture" and then hold on to it, as Nasheed also battles to? How can we make the right choices? I started this journey because I wanted to find the future but suddenly I'm feeling lost inside it. There's just too much to take in. I've not been trained for this. My education taught me to understand *this* world, not the one that's coming.

To help me make sense of it all, I'm going to meet four people who live right in the middle of the future storm of possibilities I've seen coming down the line.

And then I want to go home.

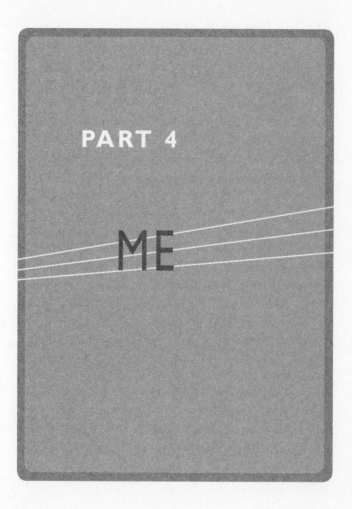

PART 4

ME

14

"Making a Road Where There Isn't One"

Are we there yet?

—Me, age 6

During my visit to Washington, D.C., to meet Google's Vint Cerf, I used my one evening in the city to tour its monuments and historical buildings. I stood outside the White House and asked a fellow tourist to take my photo, I watched children try to touch as high as they could on the great needle of the Washington Monument, and I walked the steps of the Lincoln Memorial.

On opposite walls of the memorial are two of President Lincoln's most famous speeches: The Gettysburg Address and the Second Inaugural Address. But my favorite words of his do not appear here. They were written in his annual message to Congress in 1862, when the country was in the midst of civil war. These words have come back to me now with a new resonance.

The dogmas of the quiet past are inadequate to the stormy present. The occasion is piled high with difficulty, and we must rise with the occasion. As our case is new, so we must think anew, and act anew. We must disenthrall ourselves, and then we shall save our country.

Another fan is educator Ken Robinson, who quoted these words at the 2010 TED conference.

"I love that word, 'disenthrall,'" he said. "You know what it means? That there are ideas that all of us are in thrall to, which we simply take for granted as the natural order of things; the way things are."

Sitting on the beach on Kunfunadhoo Island, I was struck by the fact that the way I think and reason is in thrall to a world that is passing. Who knew we might be able to cheat death, that we can harness the very code of life itself, that machines might be thinking and feeling, that we can manipulate matter with molecular precision, that the Internet has barely got out of the starting blocks, that data is becoming as real as flesh, that flesh and machines can merge, that we can make fuels from sunlight and air, that we can manage the climate, go off-grid, and harness the sky to remake the Earth's soils? I mean, I had a vague idea that *things were going on in laboratories*, but until I set out on this journey I hadn't comprehended what was really going on. But now I really feel it. I've seen and touched something of the world coming and met with its creators. It's been a ride. The opportunities open to us to improve the world are buzzing around my head and at the same time I worry about our collective ability to grasp them.

My personal problem is that there was nothing in my "learn-enough-facts-to-pass-the-exam" education that taught me how to deal with what I've found out. Sure I can absorb information, even make sense of how some of it fits together, but I've a strong feeling that I'm missing something. I've written this book the way school taught me—a series of essays on discrete subjects handed to you, dear reader, for comment and hopefully a good mark and a gold star. But it's not enough.

I need to make a final step and let go of ways of thinking that are so ingrained in me that I may not even be able to see what it is I need to change. The dogmas of my past are inadequate to process all I've learned. I must think and act anew. So I've come to see some people who I hope can disenthrall me—individuals known for their ability to think outside the box while living in it; people who have spotted something of the big picture. I don't want their best guess of what the future may look like, for those speculations are infinite and what will come to pass will no doubt be more strange than anyone can predict. The subjects I've investigated on my travels are not only a subset of current innovation, they are a tiny fraction of innovations we cannot yet imagine. Nobody was talking about mobile phones or data networks at the 1878 Paris Exhibition where Augustin Mouchot showcased his solar-powered ice-maker.

No, what I'm looking for now are *ways of thinking* about the emerging world that will enable me to handle and comprehend something of the march of history to come.

During the Enlightenment of the eighteenth century, reason and faith battled it out and, in many countries, eventually came to a workable compromise—a mix of secular governments, which separated religion from the business of the state, and liberal democracies that asserted the right of individuals to follow their faith unhindered. Wikipedia summarizes the popular view that the Enlightenment was "less a set of ideas than it was a set of values. At its core was a critical questioning of traditional institutions, customs, and morals, and a strong belief in rationality and science." I'm wondering what questions we need to ask of our institutions, customs and morals now. It's not that we don't have the physical tools to remake our world for the better (I've seen them), but what are the mental ones?

I need a reboot. And I'm about to get one.

My first port of call is a return trip to Boston. I'm met at the airport by Tracy Wemett, Konarka's PR liaison. Tracy has generously offered me the use of her basement room for my stay, which makes a welcome change from hotels. Still, we've got to get to her apartment alive which, given her driving, is not guaranteed.

Since I saw Tracy last, it seems I haven't been the only one to notice her maverick approach to the road. One speeding ticket too many and she's been required to take a driving education course by the State of Massachusetts. The results are reassuring.

"I was told I'm the sort of person who likes making a road where there isn't one." She pauses. "Apparently that's not good."

———

I'm back in Boston to see Ray Kurzweil, variously an inventor, guru, madman, prophet or genius, depending on who you listen to. For some he's the world's leading futurist—the man who really gets it. Others have accused him of being the ghoulish high priest of a techno-cult that has more in common with science fiction than serious analysis. For some, his ideas are just too wild or uncomfortable for them to take seriously.

I first came across Ray's ideas back in Oxford at the start of my trip. Ray, like Nick Bostrom of Oxford University's Future of Humanity Institute, is a transhumanist who intends to live for hundreds if not thousands of years. I bumped up against his ideas again investigating robotics and artificial intelligence at MIT and Cornell. Ray believes the line between humans and machines will blur and we will soon leave behind the shackles of our cruel biology and evolve, through a merger with technology, into a new species, or set of species—posthumans with powers that we cannot yet comprehend. He popped up in my conversations with Eric Drexler (mastering nanotechnology is fundamental in shaping the future Ray foresees) and Vint Cerf at Google (Ray was one of the few futurists to accurately predict the rise of the Internet). Throughout my journey I've been reading his book *The Singularity Is Near*—which talks of a soon-coming epoch, "The Singularity," "a period during which the pace of technological change will be so rapid, its impact so deep, that human life will be irreversibly transformed."[1] Depending on who you speak to, the Singularity is either pure bunkum, the next great leap forward or a dystopian apocalypse where posthumans will enact a genocide on their less developed forebears—and Ray thinks it'll happen sometime around the middle of this century. Whatever you think of the Singularity, though, there is one foundational idea underneath Ray's worldview that is hard to ignore. It's part of the reboot.

Kurzweil Technologies occupies one floor of a nondescript office block in the town of Wellesley, about fifteen miles from the center of Boston. I take a seat in the cramped reception area while I wait for the man himself. It's a slightly surreal place with a full size waxwork figure sitting opposite me, dressed in a suit and wearing a badge that announces "I am an inventor." (I later find out he's called George.) Standing forlornly in a corner behind a table displaying Ray's many books is one of Thomas Edison's early dictation machines. The walls and tables are festooned with awards too numerous to list (I estimate at least thirty) but which include the U.S. National Medal of Technology. Ray is a busy man. It's taken the entire duration of my journey to arrange a time to meet and even now he keeps me waiting for over an hour. Eventually we take a seat in his office.

He has an air of extreme calm. When I mention his critics, his defense is robust but his voice doesn't alter. He deals with every point in a cool, measured way. It's not that there's no emotion to him; it's more that he seems to have his emotions perfectly house-trained. Only when he refers to his beloved wife and family does something uncensored appear in his face—a face that evokes simultaneously both a senior academic and Billy Crystal. Watching speeches by Kurzweil is often compelling, but never because of his delivery. He'll say something that'll make your head spin in the same tone he uses to order a sandwich.* I tell him about my trip and the people I've seen. He seems satisfied with the roll call. He's collaborated with Nick Bostrom, nanotechnologist Eric Drexler is a friend, and Cynthia Breazeal appears in Ray's docu-drama film version of *The Singularity Is Near.*

Ray's journey to visionary genius/techno-prophet/crazy person (delete as appropriate) stems from his work as an inventor, in which he has a fine pedigree, as evidenced by all the awards in reception. Among other things, he invented the first machine that could scan text in any font and convert it into a computer document—a technology he applied to building a reading machine for the blind. Stevie Wonder was the first customer and became a friend, which in turn led to Ray inventing a new breed

---

*Just to be clear, Ray doesn't order sandwiches during his lectures.

of electronic synthesizers that were able to capture the nuances of real instruments.*

"Most inventions fail not because we can't get them to work," he says, "but because the timing is wrong, so I started looking at information technology trends."

What Ray found was something extraordinary—a clear, unmistakable pattern of information technologies doubling (or more) their performance while halving (or more) their cost at regular and predictable intervals. Those intervals are different depending on the specific technology—some might double in six months, some might take two years but that's not the important thing to keep in mind. What's important is that they'll take the same amount of time to double again. Most of us are familiar with this trend in the way computer-processing power has skyrocketed ("a billionfold since I was a student," he says) but this rapid doubling can be found operating in many other places too. Its significance stems from the fact that each tool we build gives us a better platform on which to build its successor. Computers for instance allow us to design more powerful computers than themselves. This phenomenon is called "autocatalysis," where the output of a process can be fed back into the process itself, spurring it on. In the opening chapter of *Whole Earth Discipline*, Stewart Brand gives a useful perspective:

> Not all technologies are autocatalytic: New discoveries don't make every technology advance faster. Progress in automobile technology and wind technology makes better cars and wind generators but not better tools for the engineering itself. The current autocatalytic technologies that goose themselves into exponential growth are infotech (including computers, communications, and artificial intelligence), biotech, and nanotech (which is blurring into biotech). What's more, they stimulate each other in a mutual catalysis that at times results in hyperexponential growth of power.[2]

---

*In a former life as a musician, I coveted the Kurzweil K2000, but not being very successful, I could never afford one.

Ray calls this phenomenon "the law of accelerating returns" and what it means, he says, is that change will come faster than we think.

Author Matt Ridley is fascinated by what he calls "the rapid, continuous and incessant change that human society experiences in a way no other animal does."[3] He puts this down to the fact that human culture allows ideas to meet and interact, something he calls "mating minds" or "when ideas have sex." It's another perspective on Robert Wright's "non-zero-sum game," the idea that trade and the sharing of culture across boundaries joins our fates together as we come to depend on one another's specialties and ingenuity.

From everything I've seen, ideas like to have sex a lot. They're not so much squeezing in the occasional fumble as going at it like bunnies. One of the things that has been getting them in the mood, according to Ray, is the power of information technology. As it increases in power, it also infiltrates all disciplines, helping us process more data, find fresh knowledge and build new tools—tools that make the next round of innovation quicker and cheaper. It helps us find ways to process genome sequences faster than we could just a few months ago, design new materials with more precision than last year's efforts, see deeper into the atom and further out into space, connect billions of us together and put processing power into every corner of the biosphere.

To understand the implications of the law of accelerating returns, you have to get your head around how potent the effect of doubling can be. Think of it this way: Let's say you travel a foot with each step you take. If you take ten steps, you'll have covered ten feet. Now imagine that instead of each step progressing one foot, you somehow double the distance you covered with the last one. So while your first step covers one foot, your second covers two and by your third step, your stride is four feet. The difference between normal-stepping and doubling-stepping is extreme and gets ever more so. As a doubling-stepper, you will cover not ten feet in your first ten steps, but more than a thousand. Instead of covering the equivalent of about one-thirtieth of an American football pitch, you'll traverse the length of more than three. And with your next step, you'll cover more than three more.

By the time you've done just twenty-seven steps you'll have traveled one hundred and thirty-four million, two hundred and seventeen thousand, seven hundred and twenty-seven feet, or just over the distance around the equator. At this rate you could walk to the sun and back in forty steps (your last step having covered 549,755,813,888 feet). One can only imagine the trousers you'd need. Meanwhile, normal-stepping-you isn't even a seventh of the way down a football pitch. Now, of course you can't step like that, but technology can.

Thinking back, I've seen numerous examples of mankind's exponential adventure on my trip: from the plummeting cost of genome sequencing to the ever improving "cost per watt" performance of solar technologies. Ray cites these examples and others. The first hundred pages of *The Singularity Is Near* almost bludgeons the reader with graph after graph, based on historical data showing exponential growth in the number of phone calls per day, cell phone subscriptions, wireless network price-performance, computers connected to the Internet, Internet bandwidth and so on. These all have a computing flavor, but Ray sees exponential growth of knowledge too, using rocketing numbers of nanotechnology patents as an example. Ray quotes example after example because he wants us to get past what he sees as an inherent prejudice in our thinking.

"Our intuition is linear and I believe that's hardwired into our brains. I have debates with sophisticated scientists all the time, including Nobel Prize winners that take a linear projection and say 'it's going to be centuries before we . . .' and 'we know so little about . . .' and here you can fill in the blank depending on their field of research. They *just love* to say that. But they're completely oblivious to the exponential growth of information technology and how it's invading one field after another, health and medicine being just the latest." He continues, "The evidence for exponential growth of information technology is very persuasive. It's the most startling thing, how smooth and predictable it is. It doesn't progress in fits and starts, which is what you'd think and what I expected. The graphs continue on a very smooth trajectory."

Ray says that what those graphs are actually measuring is innovation, creativity and competition.

"You would think that those are the least predictable aspects of human behavior and indeed each individual project is unpredictable, but the overall results are foreseeable to a high level of precision. That's because somewhere someone always makes the leap that keeps the exponential going."

I find myself nodding. I've only witnessed a tiny slice of the innovation going on around the planet but I've seen enough leaps forward to get a sense of what Ray is saying. The strongest example for me is that of Cornell's Hod Lipson, who told me about the machine he'd made that could find truths from data. "We can go from data *straight to laws*," Hod had told me. "Previously people could only go from data to predictions. So now a scientist can throw in some data, go and have a cup of coffee, come back and see fifteen different models that might explain what is going on. That saves a lot of time. Previously coming up with a predictive model could take a whole career. Now at least you can automate that so you can focus on meaning."

A machine that shortens the discovery of new knowledge by the length of *a career* is the very thing Ray is talking about: an advance in knowledge that speeds up advances in knowledge. But since I saw Hod (and rather proving Ray's point), the machine has gone one step further.

When I'd left Cornell, Hod and his team were battling to understand an equation derived from the observation of the soil bacterium *Bacillus subtilis*. The mechanical brain had come up with an equation no one could make sense of. Sure it had taken less time to come up with a law, and this was indeed allowing the team to concentrate on meaning. The only problem was, they weren't finding any. Hod wondered if it was hopeless, if the machine had given him the equivalent experience to that perceived by a dascshund being read *Hamlet*.

Now, as I near the end of my travels, I've received an excited follow-up e-mail.

"We finally understand what the equation means. It's fantastic!"

How had they found out? Long hours scratching their heads over cups of coffee and biology textbooks? No, they worked out a way to get the machine to explain it to them.

Hod and his colleagues, Michael Schmidt, Gurol Suel and Tolga Çağatay, asked the brain to conduct a new set of experiments on the *results* of the *Bacillus subtilis* investigation. These new experiments set about finding relationships between the mystery results and existing knowledge, shedding new light on the meaning of the equation. In short: robot finds out new knowledge and explains it to its own inventors. The pace of finding stuff out just sped up another notch.

Sitting in front of Ray, I'm getting just what I came for. I'm becoming disenthralled from my inclination to think linear. We must understand the power of the exponential, he urges. If we don't, progress will outrun us, and our personal decisions will be hopelessly out of step with an unfolding reality. The good news is that this ever faster rate of progress, Ray suggests, is good for the planet.

"All these concerns that we're running out of resources would be absolutely true if the law of accelerating returns didn't exist," he says. "For instance, people take current trends in the use of energy and just assume nothing's going to change, ignoring the fact that we have ten thousand times more energy that falls on the Earth from the sun every day than we are using. So if we restrict ourselves to nineteenth-century technologies, these concerns would be correct."

In other words, the law of accelerating returns should soon see a green energy revolution as solar power keeps doubling its efficiency and halving its cost, leaving fossil fuels standing. After my experiences seeing organic solar cells rolling off a printer just a few miles from here, I'm inclined toward his point of view. I've seen the kind of innovation that keeps the exponential going.

In 1999, Ray published *The Age of Spiritual Machines* in which he applied his understanding of the law of accelerating returns to make predictions, and handily he made a bunch for 2009. Critics and advocates alike have leapt on these, loudly proclaiming "Ray was right!" or "Ray was wrong!" often in the service of their own hopes and fears rather than of any reasoned analysis.

According to Ray, he made one hundred and eight predictions of which eighty-nine are correct, thirteen are "essentially correct," three are partially correct and two are ten years off. Just one is wrong, he claims, but that was tongue in cheek anyway, as it predicted that the two most popular choices for intelligent computer assistance would take the form of "Maggie, who claims to be a waitress in a Harvard Square café, and Michelle, a stripper from New Orleans." I counted his predictions slightly differently and concluded he got nearly two-thirds right, including forecasts about how we'd interact with media, the power of computing, the rise of networks, the growth of wireless connections, the role of technology in tackling disabilities, advances in nanotechnology, the issue of privacy as a political hot potato and improvements in medicine. Of the rest, about half I put in the "sort of right" category, by which I mean that the prediction has come true, but not quite in the form Ray describes it, or it's something that looks like it's coming soon (for instance, "the greatest threat to national security comes from bioengineered weapons"). The remainder are "wrong" but only in that Ray was optimistic on the time frame. For instance, critics tend to leap on his prediction that "translating telephone technology (where you speak English and your Japanese friend hears Japanese, and vice versa) is commonly used," which hasn't happened, but to cite this one example as a way to suggest "Kurzweil is wrong" is a bit like saying Luciano Pavarotti was a rubbish opera singer because he once hit a bum note. One also has to bear in mind that exponential growth may look small at the outset. Doubling from 0.1 percent to 0.2 percent doesn't look like much, but when things start leaping from a thousand percent to two thousand, they seem to "come out of nowhere" even though the exponential trend has stayed constant. Translating telephone technology could arrive with just such a bang at some point in the near future.

I've spent my journey seeing much of what Ray forecast in 1999 coming true. You can't hide the fact that so far he's been right a lot more often than he's been wrong. By anybody's reckoning that's prediction nirvana and a skill any investor would soon crow about.

Oh, Ray's latest venture? A hedge fund.

If Ray's right about the law of accelerating returns, and I think when it comes to infotech, nanotech and biotech he is, this means that when I decided to investigate the future of my lifetime I was being hopelessly overoptimistic. What I've actually been exploring is the next ten or twenty years at best. Everything I've talked about on my journey—from Eric Drexler's nanofactory, to the biological immortality of Nick Bostrom, to George Church's "personal genomics," to Cynthia Breazeal's sociable robots—is coming quicker than we expect, along with synthetic fuels and a solar power revolution. By the time I'm one hundred, the ever increasing rate of change will have made whatever is coming literally indescribable today. Trying to write a book about that would truly be an exercise in science fiction.

It's talk of such rapid technological evolution that brings Ray into conflict with other scientists and commentators who think he's not so much running before he can walk as getting into the cockpit of a jet fighter straight from the crib. For Ray, though, that's kind of the point—crib to jet fighter is really just a few doublings, the law of accelerating returns in action. In fact, Ray points out that sometimes the rate of doubling can double itself, creating the "hyper-exponential growth" Stewart Brand references. Others are unconvinced. They see Ray the same way the State of Massachusetts sees Tracy—he makes a road where there isn't one. Douglas Hofstadter is one critic. Now a cognitive scientist at Indiana University, he more famously authored *Gödel, Escher, Bach*—an attempt to explain how consciousness can arise from a system, even though the system's component parts aren't individually conscious.* Hofstadter told

---

*This is a key area of study for Ray too, because it is through "reverse engineering" the human brain that he believes we'll be able to unlock the mechanisms of the mind, replicate them in machines and so free ourselves from the biological limitations of our brains.

American Scientist he thought Ray's ideas were like a blend of "very good food and some dog excrement," which made it hard to untangle the "rubbish" ideas from the good ones.[4]

That's gotta hurt.

My own feeling is that Ray plays with ideas that touch on some of our deepest hopes and fears, and that's why people react.

"Take death," says Ray, which is a hell of a way to open a sentence. "We've spent thousands of years rationalizing death. There was no argument you could make for defeating it until fairly recently. People still disagree but you can make a very credible argument that we can defeat death, not in the mystical way that traditional religions give us by saying death is ennobling and we live forever on the other side, but by using technology. All of our institutions are going to have to be rethought. Marriage, for instance, is based on a premise that people aren't going to live that long."

There you go: death, religion and sex in one paragraph. Ray is the Tennessee Williams of futurology. It takes enough effort for our linear brains to get themselves around just one or two of these ideas and then Ray comes along and says, "Belt up, things are going way faster than you thought, and by the way that means I don't intend to die. Would you like to transcend your biology with me and embrace the next stage of techno-human evolution? Hurry now." It's no wonder some people find Ray just *too difficult* to engage with and it's easy to understand why. In my research, I've found myself reacting strongly to his ideas too.

But what I am taking away from him is the understanding that coming to grips with the exponential growth of technology is a key part of my personal reboot. I want to ask Ray a wider question, though.

"Have you got any graphs that clearly show an exponential growth in our ability to make sense of the great philosophical questions: 'What is life?' 'What is consciousness?' Have we seen the law of accelerating returns in our understanding of these questions? Is the growth of wisdom keeping pace with our technology?"

Ray falls silent, contemplating, and then deflects slightly, telling me that he thinks machines are getting closer to what we would call alive and conscious. But I push back. That's not quite the question I asked.

"Well I think these philosophical discussions are now becoming very pressing. For instance, at what point do we have to consider a software program as having human rights? We're thinking more about it than we ever did before."

It's an echo of something MITs Cynthia Breazeal said to me—that her research into robotics "pushes us to ask those questions, 'What is life?' 'What is mind?'" She'd told me she thought the answers were "decades off." If Ray is right, technology will force our hand far sooner.

I'm staying in Boston for the next part of my reboot: a meeting with Juan Enriquez, one of the world's leading authorities on the impact of technology (particularly biotechnology) on our economies and politics. Our conversation takes place on the fifteenth floor of Boston's Prudential Tower, in the offices of Excel Venture Management, a venture capital and investment firm Enriquez runs when he's not lecturing around the world, writing books or serving on the boards of numerous businesses and academic organizations.

Juan is a reserved and gentle man in person, reminding me of Kelsey Grammer (famous for playing radio psychiatrist "Frasier") although with more hair, a slightly slimmer face and a beard. He describes his life as "a series of strange accidents," which is a rather self-effacing way of summarizing an impressively eclectic powerhouse of a CV. Those "accidents" arguably started rolling off the conveyor belt when, as a young man living in Mexico, Juan walked into his parents' room and said, "I'm not learning enough here, so I'm going to go to school in the U.S."

"I applied late, I had no idea it was hard to get into these places and even though I spoke English (my mother's American) I'd never studied and written in English. I have no idea why I was admitted. I mean during the placement exam I was asked to write a paragraph and I asked, 'What's a paragraph?'"

He describes feeling "utterly stupid" for his first semester but obviously caught up quickly and maintained that accelerated intellectual velocity. He was admitted to Harvard to study government and economics, after which he returned home to "change Mexico"—a childhood ambition born out of a belief that his homeland too readily disadvantaged those not in the ruling class. "I always thought I would work in and change Mexico. I was bothered by the poverty I saw there." He became the country's youngest budget director ever, and then returned to Harvard before being offered "a dream job" back in Mexico as head of the Urban Development Corporation. So far, so impressive (especially when you consider that during his time in Mexico, Juan was also part of the team that negotiated peace with the Zapatista rebels). And then Juan discovered something more important, a revolution that would not only affect Mexico but the entire world, and all because of some lonely-looking geeky guy at a New Year's Eve party.

He recalls, "There was this guy sitting over on a corner table by himself and I think 'Poor bastard, it's New Year's.' I walk over, sit down and talk to him for the rest of the night. By the end of the evening we decided to sail across the Atlantic together in two weeks. By the end of that trip, I had decided that I was going to change my entire career and learn biology."

The "poor bastard" in question was Craig Venter, famous for sequencing the human genome. Juan recalls, "That conversation was so interesting. All of a sudden I thought, 'I want to learn about this.' I wondered, 'Who gets affected by this stuff? What does it do? Why does it matter?" The sea voyage they took went in search of marine genomes, uncovering myriad new species of microorganisms. Of the forty-one million genes in public databases, forty million come from this single expedition. On the trip Enriquez told *Wired,* "I gave up a lot for this. I canceled a meeting with [President] Bush and blew off a couple of foreign ministers." Well, there's an image. "But what could be more exciting than sailing around the world, discovering thousands of new species?"[5]

In his book, *As the Future Catches You,* Juan writes:

Your future, that of your children, and that of your country depend on understanding a global economy driven by technology. Understanding code, particularly genetic code, is today's most powerful technology.[6]

Embarking on the marine voyage, he said, "If you do not perceive the possibilities in this shift, if you say *no* instead of *yes*, you will be left in the past. There will be whole societies who end up serving mai tais on the beach because they don't understand this." Juan is very clear that new technologies quickly alter the fate of nations. The rules are changing.

"It took me a damn long time to figure out. It's Darwin. It's the ability to adapt and adopt. It's not the most powerful who survive, it's those who best adapt to change. One of the most important things that people keep forgetting about America, and the reason why I think America truly became a world power, is because so many of the founders were adamant about education and science. Just look at Franklin, or Jefferson, and you'll see people deeply committed to critical thinking and education. There was a huge tradition of science and technology education, freedom of inquiry, and that's powered this country in an extraordinary way."

Thinking back to my conversation with Ray Kurzweil, I give voice to the fear many feel when they hear about the implications of the infotech, biotech and nanotech revolutions currently "goosing themselves into exponential growth"—that technology is running too fast for us to make sense of.

"If you look at a lot of the things we're building, they're scary as hell to some people. You talk about programming cells or sentient robots or evolving the species using technology—that is profoundly disturbing to some people because this stuff is very powerful. It upends industries, it changes how long we live, it changes what our kids may look like. I look at that stuff and say, 'Okay, it allows people who couldn't have children to have children, we're going to do away with some of the diseases, and so on.' Other people look at that in absolute horror. They say, 'Stop the world. This isn't natural. This isn't what God ordered. I want to get off.' They're looking for an element of stability and certainty. This desire tends to manifest most during the periods of fastest change, like now. You want something to hold on to. And if you're not part of that ride, if you don't think you can play in that game, then you get an anti-intellectual counterpoint."

In short, some of us are running scared, which, according to Juan, is about the worst thing you can do. Mark Bedau, editor of MIT's *Artificial Life* and a philosopher with a particular interest in synthetic biology, told me, "Change will happen and we can either try to influence it in a constructive way, or we can try to stop it from happening, or we can ignore it. Trying to stop it from happening is, I think, futile. Ignoring it seems irresponsible."

You can stop the advance of knowledge in some places for a while if you're brutally draconian or conservative, but not for long. Beyond that, lost knowledge is soon rediscovered, and the more technology allows autonomy of the individual (from wireless Internet access to the world's knowledge via mobile devices, to power independence through solar technology) the harder it becomes to suppress our advances, good or bad. Hoping to roll back the march of human culture is about as futile as trying to reconstruct the frog your younger brother just put in the blender. The frog is dead. Deal with it. As Septimus Hodge says in Tom Stoppard's *Arcadia*, "You do not suppose, my lady, that if all of Archimedes had been hiding in the great library of Alexandria, we would be at a loss for a corkscrew?"

Juan isn't worried about our self-directed evolution. "The notion of evolving into something else is terrifying until you consider the question 'Are popular radio jocks the be all and end all of evolution?' If that's all Mother Nature wrote, *then* I'm scared. I look at this stuff and say, 'If my kids could live two hundred years with a good quality of life, if they could see a lot further than I could, if they could regrow their joints, if they could hear a lot better than I can, if they could have brains that were fifty times as powerful as mine? Good for them. Cool. I'd rather things carry on.'"

So, the second part of my reboot is getting used to the idea that evolution is far from over. It's something I'd kind of understood when I met George Church, and Ray Kurzweil had discussed it too. Juan is ramming it home and he needs to. I've put up a mental barrier to this somewhere in my head. Like musical theater, I get it in abstract but want nothing to do with it personally, fearing if I do I'll lose my grip on something essentially human. I have to get my head around the fact that as a *Homo sapiens* I am only a staging post on the way to another form of life that's coming down the line. Attempting to hold on to something "essentially human" by trying to fight against the very thing humans do best— evolve through culture and technology—is a contradictory and ultimately futile course of action.

But can our morals keep up with all this change? Einstein famously said, "It has become appallingly obvious that our technology has exceeded our humanity."

Juan makes an interesting observation.

"Every civilization has, to a greater or lesser extent, some religious moral background. There has to be some evolutionary advantage to having that kind of moral backbone and that kind of belief system, and I think it's because it traces how you move from a hunter-gatherer society, where everybody knows each other and watches each other all day, into a town, into a city, into an empire. . . . And just like most animals, almost every religion and God has gone extinct. The interesting question is, which ones survive, how do they survive and how do those moral backbones evolve? We have to ask ourselves, What does a moral ethical framework look like if we start to alter fundamental characteristics of what we consider human?"

The way I interpret this is that if there is an evolutionary advantage to having a moral framework or religion, it had better evolve with you. This, I think, doesn't mean watering down the essential need for compassion, the "golden rule" of treating others as you would like to be treated, but it does mean we need to work out how to continually keep that ambition central to our societies in a rapidly changing world. There's plenty of work for theologians yet. Indeed, you could argue we've never needed them more, although they, like the rest of us, may have to adjust to a new time frame. Like Ray said, those philosophical questions we've always battled with are going to demand some answers soon.

One thing is for sure. The future won't be a smooth ride.

"Things evolve at different times at different paces, people make different choices and that's one of the reasons countries disappear," says Juan. "There really are consequences to your choices. If you choose to shut your doors and not follow technology, you will vaporize your sovereignty. So, there are galactically stupid policies as far as individual countries are concerned. The future of the species worries me a lot less."

My third stop in search of the reboot returns me to California to meet a man called John Seely Brown (or "JSB" as he's often referred to). In a former life he was the director of the Palo Alto Research Center, a hotbed of innovation that is widely attributed with inventing laser printing and graphical interfaces for computers. Now in his late sixties (though he looks a good ten years younger), he works independently, helping organizations adopt and adapt to new technologies. It's on the matter of how our organizations and society will have to change that I've come to talk to him. He dubs himself the "Chief of Confusion" and sees his job as "helping people ask the right questions." He's particularly interested in what he calls "radical innovation." His website tells me he is "part scientist, part artist, and part strategist" as well as a keen motorcyclist. Vint Cerf told me, "John is among the best thinkers in the United States and probably in the world. He's just a brilliant, modest, very, very deep guy." After all this, if someone told me he also has a part-time role as Batman, I'm not sure I'd be surprised.

I meet John at his house in the appropriately named district of Professorville, Palo Alto, on a gorgeous California day. As I'd done with Ray and Juan, I briefly recount my trip and tell him I'm looking for some ways of thinking to make sense of it all. He listens intently and I get the impression that every word I'm saying is being catalogued and cross-referenced as I say them. He has the manner of an owl set into the face of a kindly bear and reminds me of my best teachers at school—the ones that always managed to interest you with their subject but never pretended to be friends with you. He imparts knowledge in the same way Keith Richards imparts riffs, with a kind of blistering efficiency and little worry about what this might do to your head.

John believes we're "moving into a world of constant flux, rather than a world of basically slow, punctuated evolution. If you take this to the extreme, it says nearly every social and technical and business infrastructure we have can't survive."

Well, we are near the end of the book. What did you expect?

15

"Future Shock"

When you are through changing, you are through.

—*Bruce Barton*

The Industrial Revolution changed everything.

It changed the way we organized society, it changed the way we educated ourselves, it changed the idea of work and it changed the way we did business. We built an infrastructure the like of which had never been seen before. Railways, roads, sewers, waterways, ports. These were new technologies. Today we don't think of the road or the sewer as a technology. But they are. Vint Cerf told me, "If you grow up with a technology, it's not technology. It's just there."

In 2009, John Seely Brown told a crowd of Silicon Valley business leaders how deep the Industrial Revolution was embedded in their high-tech businesses. What he said isn't immediately obvious but it is profound.

"The structure and architecture of the firm reflects the structure and architecture of the *infrastructure* on which the firms are built," he said.[1]

In other words, society is built *on top of* our infrastructure, not the other way round. Seely is saying that our roads and schools shape us far more than we shape our roads and schools.

As educator and Abraham Lincoln fan Ken Robinson points out, "There were no systems of public education around the world before the nineteenth century. They all came into being to meet the needs of industrialism."

We built our school systems on top of the Industrial Revolution. We didn't build the Industrial Revolution because we'd sent people to school.

The Industrial Revolution is still with us. The infrastructure grew rapidly and then reached a plateau. Cars are not radically different from their counterparts a generation ago, nor are trains, aircraft, roads, railway tracks, airports, the judiciary, schools, universities or our systems of government. "Our infrastructure has been stable for a shockingly long period of time," says John, "and we have now built institutions on top of it that expect that kind of stability."

"After everything I've seen I've a feeling that's about to change," I say.

"I think right now we are experiencing something we've never experienced in the history of civilization. All past infrastructures have unfolded slowly at first, until they reached a critical mass which then sparked explosive expansion and adoption. Finally things level off and stabilize for decades at a time, often seventy to a hundred years. What's interesting for me is that the infrastructure we're moving into is an exponentially increasing infrastructure, because technology is exponentially increasing."

"You agree with Ray Kurzweil?"

"I think Ray talks about this to the extreme and he doesn't account for the fact that technology is constrained by society. If you ignore society, it's easy to take the exponential prediction too far. But yes, I agree that our digital technology has exponential growth. The question is, how is society going to interact with that?"

By way of example John talks about the introduction of electricity.

"Electricity was a new technology, but integrating it with society brought social and organizational innovations too. You couldn't give electricity to the people if you didn't agree on standards and ways to distribute the power, so in the U.S. we started out in a highly fragmented way with each city or county building their own isolated system, but eventually a country-wide standard emerged leading to our nationwide electric grid. That was a huge social and business innovation. All infrastructure is social-technical."

I'm thinking back to my conversations with Rick Hess about the off-grid potential of solar power, how the days of an electricity system dominated by centrally generated power may be coming to an end—and wonder, if it comes to pass, how society will make sense of that.

The current model of electricity generation and distribution is just a tiny example of what John calls "scalable efficiency" where organizations get bigger in order to drive down costs. These organizations have to be able to predict demand, hoping they are producing the right amount to match it. Too much and their costs will be out of kilter with their income, too little and they miss an opportunity, upsetting the end user. This book is another example of a "scalable efficiency" operation. Print too many copies and the publisher will be sitting on surplus stock, publish too few and sales are lost.

But all this is changing, believes John. Digital infrastructures allow us to usurp these old ways of working. If you're reading this as an e-book, for instance, then there were almost no physical costs of production and there was no need for the publisher to ponder how many copies of the book to manufacture and ship. They just sent you a file. The reproduction costs are roughly that of sending an e-mail. Everyone wins. As a customer you are never frustrated by a book being out of stock, and the e-version is cheaper too. As a publisher you're never at risk of judging demand wrongly and making too many or too few copies. The author is happy too, because this reduction in risk and production costs translates into a higher royalty. The book's price becomes a closer reflection of its value to you as an interesting read, rather than as a function of the physical costs of production. Another option is print-on-demand books, where onsite digital printing and binding machines create a physical book in response to an individual order. Thinking back to my conversation with Eric Drexler, his nanofactory may one day be the ultimate expression of the new digital infrastructure that society is shifting to. Maybe you *are* reading this in the future, having dialed up the manuscript from an online archive that your nanofactory (or whatever home fabrication system you have) made into a physical object. If so, tell your machine I knew its father.

"I don't know if you've sensed it," says John, "but innovation is just again taking a huge punctuated jump. All of a sudden we are finding completely new ways to create things. These creations are being enabled by the mass of digital infrastructures enabling people to do stuff that would have been impossible before."

John tells me of a project he's involved with to create a new kind of nanoparticle that mimics the properties of platinum, an increasingly rare resource.

"This is the sort of thing that only a few years ago you'd have had to ask a national lab to look at and commit many millions of dollars and masses of manpower to. This start-up is doing it with ten young engineers and physicists in a garage where they grab the computational resources they need via the Internet, just buying massive computing power by the hour incredibly cheaply. They're not even starting with combinatorial design (where you try out thousands of compounds to see if you can get something that works) but designing the thing from first principles—i.e., *from the bottom up*. I mean, who the fuck would've tried to go and build a nanoparticle in a garage? It's not something that would have made any sense before. Now it does."

As John is speaking, I can see two ideas I've encountered on my journey interacting. Vint Cerf's "Internet of things" has taken on a whole new power, now linked to Eric Drexler's ambition of atomically precise design and manufacture. It's Matt Ridley's notion of ideas having sex again. Internet, meet nanotechnology: I'm sure you'll get along famously.

"Perhaps more importantly," says John, leaning forward, "you now only need a *tiny* amount of money. What we're seeing are very small, very agile companies that are capable of getting off the ground in a way that was previously unthinkable. They don't have an overbearing investor breathing down their backs. They can fund themselves through friends and family."

I'm reminded of how things are likely to get cheaper as information technology infiltrates manufacturing—the post-capitalist world I'd discussed with Eric Drexler.

John's speaking with a heartfelt urgency now. "The old-age companies don't know why they have to run faster in order to lose more slowly," he says, laughing. "All the practices of those companies are exactly the practices that keep you from being able to engage in the world of fast-paced innovation. They have routines and beliefs built on the assumptions of stability. Almost any company that's more than twenty years old isn't built right for this. In fact, I would argue that companies that are five years old aren't ready either."

John and his coauthors of *The Power of Pull* offer a stark warning: "The world is transforming around you. The truth is, the thing you did to get there will no longer work to keep you there." This is something they call "The Big Shift," which is already affecting every aspect of society. In particular, educational institutions are grappling to make sense of "the quickly changing learning needs of students."[2] As Ken Robinson said in 2006, "If you think of it, children starting school this year will be retiring in 2065. Nobody has a clue what the world will look like in five years' time. And yet we're meant to be educating them for it."[3]

I recount something Vicki Buck said to me as we sailed in Queen Charlotte Sound in New Zealand.

"The fear I have is that the education system takes really creative, wonderfully imaginative little kids and seeks to provide conformity and obedience and discipline," she'd said. "I find that scary in a time when those kids are probably going to have to create their own jobs."

John agrees. "Education itself is delivered using a factory model. You build up an inventory of skills. You were told these would be useful at some point. What starts to happen when you move into a world where the life of a skill is five years instead of sixty? Our current educational institutions make almost no sense anymore."

It's not just education.

"Governments are racing to make sense of a world where social and economic changes far outstrip the ability of legislatures and even dictators to keep control," he says. "It's a different game now, and many of us have yet to learn how to play it."

Suddenly the world I've grown up in is starting to look like a relic, and I realize that almost nothing in *my* education has prepared me for a future built on an exponentially growing information infrastructure instead of a static industrial one. I've been trained to work for organizations that centralize production and I've been spat out of an industrial educational system to fit their needs. What happens if the world shifts toward a networked infrastructure of locally manufactured abundance?

Rather than feeling elated by what John is saying, I'm slightly terrified. It's not because I think any of these innovations are a bad idea; it's because quite simply I don't feel ready. It all seems to be coming too fast. I'm almost expecting that by the time I walk out of John's office I'll find the world already transformed into some futuristic society that I've no idea how to operate in. It seems I have the first mild onset of a phenomenon identified by sociologist Alvin Toffler: "future shock," which can be summarized as too much changing too quickly for you to get your head around. Part of the reboot, then, is finding a way to deal with change as the norm and not the exception, realizing perhaps that society, having enjoyed a period of relative stability since the Industrial Revolution is, in my lifetime, going to see a radical shift. If the infrastructure we're shifting toward really is exponential in nature, then I need to learn to let go of ideas on a regular basis.

Have you ever agonized about breaking up with someone? You know, that terrible soul-searching period where you know it's not working but wonder whether you've really given them a chance, or put enough effort in, or you worry that the moment they're gone, you'll realize it's madness to be without them? But if it is the right thing to do, you feel sad about the trauma but lighter somehow, like a new window of opportunity has opened up somewhere inside you. You'll take what you learned from the last relationship and make the next one better, you know it.

I think this is how I'm feeling about the industrial capitalist society I've grown up in. I'm scared to break up with it. We've been together so long.

"There may be a whole new system where you're defined more and more by who you are and not by what you own, by what you've created and shared, and what other people have then built on," says John. "If you look at what's happening in nanotechnology, in computing, in medicine, you see increasing acceleration of capacity at lower and lower cost—and you're also seeing this trend whereby you are valued not by how much you own but by how much you innovate. The skills that will help you prosper will be based around a questing disposition. That is not typical. That is not considered a twentieth-century skill, but it may be considered more and more a twenty-first-century skill."

I'm not scared now, but excited. This breakup is starting to feel right.

"We're getting to a place now where we can all live with perhaps a lot less money than we used to, which means you'll see more and more people breaking free of some of the nineteenth- and twentieth-century notions of capitalism as the sole basis of being. But if the world is now accelerating like this, your degree of comfort depends on your ability to embrace change—not flee change."

"A lot of people and organizations are going to find that difficult, surely?" I say.

"A fundamental clash does exist and the question is, how do you start changing the shape of institutions to embrace this? The real challenge of innovation today is not technological innovation, it's institutional innovation. We have to start inventing new types of institutions that can stay in step with the information age."

I find this a powerful thought. I've spent my journey looking at technological innovations (from new ways to fence cattle through to personal genomics) but John is saying it will be how we organize our society in response to these ideas that will be the real innovation. We have to break up with the Industrial Revolution and finally form a proper relationship with the Information Age: one based on the continual growth of ideas rather than bank balances.

John tells a story. "I was recently asked to suggest the innovation in the last three hundred years that had generated more wealth for mankind than anything else. I knew why I'd been asked—they presupposed I was going to talk about the microprocessor and I disappointed them dramatically. I said, 'The innovation that's generated more wealth than anything is the limited liability corporation because that enabled you to accumulate, invest and leverage your wealth but with limited liability.' That is what actually unleashed the power of the nineteenth and twentieth centuries, that one innovation. Isn't it curious when you talk about innovation today, you hear *almost nothing* about institutional innovation?"

The third part of the reboot is complete. Whereas power used to be handed down from above, it's increasingly going to come within reach of us all. Hierarchies are on the way out, networks are on the way in. Don't look to get promoted, look to get distributed—as an individual, as an organization, as a nation.

Marketing guru Seth Godin has said:

> We started with the factory idea. That you could change the whole world if you had an efficient factory that could churn out change. We then went to the TV idea. That said if you had a big enough mouthpiece, if you could get on TV enough times, if you could buy enough ads, you could win. And now we're in this new model of leadership, where the way we make change is not by using money, or power to lever a system, but by leading.[4]

I like the idea that your personal worth in our coming society will be in how many people you inspire, not how much money you have. The rich will no longer be rich because of what they own, and it's going to get increasingly tough for those surviving on inherited wealth as the century gets fully under steam. The best inheritance you can get, or the most useful thing you can acquire, is what John calls "the questing disposition." The value of money, ironically, is decreasing. The value of a good question is rising.

So, what do you want to ask?

The final destination in my search for a reboot is a set of offices high above the teeming streets of New York. I'm meeting a man who, by his own admission, used to be mildly delusional.

In February 2002, in front of an audience of some of the world's leading thinkers, he said, "I think that I may have believed unconsciously then that I was kind of a business hero. I had this company that I'd spent fifteen years building. . . . It had recently gone public, and the market said that it was apparently worth two billion dollars, a number I didn't really understand."[5]

But then it all went horribly wrong as the dot-com bubble burst.

"I entered eighteen months of business hell. I watched everything that I'd built crumbling. . . . [It] took eight years of blood, sweat, and tears to reach three hundred and fifty employees—something which I was very proud of . . . In one day we laid off three hundred and fifty people, and before the bloodshed was finished, a thousand people had lost their jobs from my companies. I felt sick. I watched my own net worth falling by about a million dollars a day, every day, for eighteen months. And, worse than that, far worse than that, my sense of self-worth was kind of evaporating. I was going around with this big sign on my forehead: 'Loser.' And I think what disgusts me more than anything, looking back, is how the hell did I let my personal happiness get so tied up with this business thing?"

Today Chris Anderson is a lot happier.

He managed to save his companies but he'd had enough. With time on his hands he started reading again, "discovering, *holy crap!* There's so much amazing new thinking out there. That was a kick." I know what he means. Before I decided I actually wanted to answer the question "what's next?" I was on the same treadmill: too busy to look up to realize that the story we're so used to being told wasn't necessarily the only game in town. Chris's revelation led him to become the "curator" of the TED talks, a growing set of conferences and an Internet phenomenon. TED is a platform for "people who can offer a lens through which to see the world in a different way," he tells me as I take a seat. He'd lost interest in money as a currency and had become interested in the currency of ideas. He had rediscovered what John Seely Brown calls "the questing disposition."

Chris has a boyish face that's somehow beatific. He's just past fifty but seems ageless. From a distance he could easily pass for a man in his early twenties. There's an openness to him that is reflected in his posture, the soft tones of his voice, the way he's happy to be approached by ideas and happy to approach them. He's thoughtful without being serious, sharp without being aggressive, friendly with no need to make friends. My first impression is almost that of a blank canvas, but as our talk progresses I change my mind. He's not a blank canvas, he's a constantly changing one, repainting himself willingly when new knowledge comes his way.

Every year, Chris and his team gather together some of the world's best disenthrallers and give them eighteen minutes apiece to tell the rest of the world how they see things. In 2006, the talks started going on online and they've become a phenomenon. Hundreds of millions of downloads pay testament to a hunger for new ways of looking at the world. TED does no advertising, it's a word-of-mouth thing and it's easy to see why. The ideas you'll find at TED are by turns so mind-bending, hopeful, scary and entertaining that they demand to be shared. There isn't a film out of Hollywood I've seen that has twisted my brain and entertained me as much as some of the ideas paraded at TED.

Mixing the most radical ideas with a short format means speakers need to hone their presentations, and because the Internet audience is millions there is no room for academic long-windedness. In 2007, statistician Hans Rosling (whose thoughts on population I'd considered while walking the packed streets of Malé) delivered a blistering attack on the concept of "developed" versus "developing" nations . . . and then he swallowed a sword, because sword swallowing is "a cultural expression that for thousands of years has inspired human beings to think beyond the obvious." Of course, Hans, it's an entirely natural way to end a presentation on statistics. But that's TED.

On ted.com you can see Juan Enriquez talk about the coming age of genomics, Ray Kurzweil summarize his law of accelerating returns and Hod Lipson demonstrate his self-aware robots. You can see Ken Robinson ask us to reevaluate what we need from our educational systems, Steven Pinker tell you the world is less violent and then watch Robert Wright explain why that might be, along with a host of other mind-shifting presentations that make you see things from a different angle.

I've come to see Chris because in putting together the TED talks he is probably assailed by more new ideas than pretty much anyone on the planet—and he has to make sense of them somehow. I ask him how he synthesizes everything.

"Well, like everyone else I'm on a journey," he says. "But there's a very boring view of the world which is that 'things happen' and basically history is one thing after another, the idea that you really can't say much about the future, other than it's probably not going to be as good as you think. But there are bigger trends at work, like Ray Kurzweil's 'law of accelerating returns': the idea that it's in the nature of ideas themselves to build on one another."

Like John Seely Brown, Chris has seen a quickening of the pace of innovation in recent years.

"The acceleration of knowledge and ideas made possible by the fact that humanity is connected for the first time is vast," he says. "This is happening in hundreds of areas of human endeavor."

By example he talks about how the quality of contributions to TED has been improving since the conference went online.

"Speakers are looking at what other speakers are doing and are putting in far more preparation time than they ever used to." (How many, I wonder, are taking sword-swallowing lessons?) "It's a big kick for me to see these people competing. I've started to call it 'crowd accelerated innovation' and it's not something exclusive to TED."

Another example is the Legion of Extraordinary Dancers, who had come up in my conversation with Vint Cerf—a dance troupe built from kids innovating on each other's moves (continually posted on YouTube) and who appeared at the 2010 TED conference in Long Beach, California.

Chris thinks that rather than technology separating us from our humanity, it could help us to rediscover it. After all, what could be more human than the Legion of Extraordinary Dancers? Kids with diverse backgrounds from across the world, innovating and collaborating to bring a new dimension to an art form as old as society, using technology to help them express themselves and innovate physically with their bodies, to meet, to collaborate, to just *dance*—and then show the world.

Look what we did. Here is something of the exponential growth in wisdom, community and understanding I was looking for to go with Ray Kurzweil's accelerating technologies.

"We're pretty sure the biggest growth in the TED audience is people in their teens and twenties," says Chris. "A lot of people fear that the generation coming in are made up of noncurious and not very hardworking slackers, but there's a lot of evidence contrary to that. Increasing interconnectedness is changing the psychology of the generation coming through; they really believe that they can make a difference to the world."

"So it's not just that they're curious. They also have a sense they can do something?"

"Exactly. I think the sense of your ability to play a role in an interconnected world is changing. And the reason they believe that is because it's true."

"Does that make you optimistic?" I ask.

"I don't know that the future's going to be better," says Chris. "But I think there's a very good chance that it will be and I think there is something everyone can do to further increase that chance. Of course, the future might well be truly horrible. I think it's all to play for and I think everyone of sound mind and conscience should be in the game, trying to shape it in the right way."

As I'm leaving, Chris asks about my solar bag. I explain the technology and how Konarka was founded on the principle of taking cheap power to areas without electricity grids.

"That's superb!" he exclaims. "Fantastic. Fabulous. There are so many things like that that can get you really excited."

At the door he says, "You know the title for the next TED conference?"

"No," I reply.

"We've decided to call it 'And now the good news.'"

I take a walk through Manhattan, and find myself sitting on the steps of the New York Public Library watching the crowds go by. It's a baking hot day. Did I get what I came for, the final part of my reboot? There's something in my conversation with Chris that's bashing around inside my head but hasn't settled.

Often when I see crowds of people I'm struck by the fact that inside every single head there is a separate personality, a separate consciousness, a unique story. I sometimes try to imagine what that story might be. But today I find myself thinking about our collective story and who tells it. We all have our individual tales but who looks after the story of human-ity? We have plenty of biographers (they're called historians), but who is scriptwriting the next episode? Chris talked about rejecting the "very bor-ing view of the world which is that 'things happen' and basically history is one thing after another," the idea that "you really can't say much about the future, other than it's probably not going to be as good as you think."

I'm thinking about the world I've grown up in, recalling the story of the future I've been used to hearing in the discourse of politics and its reflection in the media. Despite everything I've seen on my journey, I feel my heart sinking. My perception of the story we seem to be telling ourselves is: "Life happens to you, the future is not going to be very good (especially if you vote for that guy), it was better in the old days, you've got to look after yourself, the world is violent and unsafe, your job is at risk, the generation below you are feral and dangerous, things are chang-ing too fast and you can't trust those scientists/new-agers/left-wingers/right-wingers/nerds/geeks/religious people/atheists/football fans/the rich/the poor/what you eat/your neighbor. You are alone. Make the best of it. Vote for me. Buy my paper. I understand."

What's worse is that it's a story that seems pretty constant. I can't remem-ber a time without it and, as I think about it, I can feel its fingers reach deep into my soul. I find myself chastizing the optimism of those I've been visiting, and my own naïveté in being taken along with it.

And then the final part of the reboot goes off in my head like a firework.

16

Hello, and Thanks for All the Fish.

Following the light of the sun, we left the Old World.

—*Christopher Columbus*

've traveled more than sixty thousand miles across four continents, talked to more than thirty geniuses, met four robots and had two terrible conversations with computers. I've contemplated immortality, the end of capitalism and a new age of human evolution. In the process, I've attended an underwater cabinet meeting, helped invent one cocktail, seen spaceships, been insulted in the Outback, made a brace of new friends and had one near-death experience (thank you, Tracy). I'm not the same person I was when I started.

Now I'm sitting in my local park overlooking London. Just behind me is the spot where in 1795 the British Admiralty erected an optical telegraph station to pass signals down the line between coast and capital. Communications have come a long way since 1795. On my lap is a computer, battered and grubby from long hours on the road. Using my mobile phone as a wireless modem I am surfing the Internet. In particular I am looking at my "genetic profile" having just logged on to the website of California-based 23andMe, one of a new breed of personal genomics companies.

Several weeks ago, the company sent me a plastic tube, which I filled with saliva and returned to its laboratories. From this the company extracted cheek cells, out of which they stripped my DNA to be duplicated many times over. These synthetic copies of my DNA were then chopped up and applied to a "DNA chip," a glass slide with millions of DNA "probes" on its surface. These DNA probes are mirror images of locations on the human genome that are known sites of genetic pointers that can indicate your susceptibility to particular diseases, the chances of you exhibiting a particular trait or give clues to your ancestry. Because DNA's component parts love to snap together in predetermined ways, copies of my own genetic code would have latched on to these probes with varying degrees of connectedness, letting the company know which markers I possess and allowing them to suggest what these might mean for my health. Because our understanding of the interplay of our genes and environment is still evolving, 23andMe attaches confidence ratings to each finding (the higher the rating, the more secure they feel in their analysis).

Because I have one genetic marker that a 2007 German study suggests is linked to Tourette's syndrome,[1] 23andMe let me know I *might* have an elevated chance of the condition, although they give this a confidence rating of one (out of four). In the two-star category there are potential elevated risks of "essential tremor," "Hashimoto's thyroiditis" and "Sjögren's syndrome" (the madness brought on from not being able to pronounce the name of the syndrome you have*). The company gives a confidence rating of three to its analysis that I have higher-than-average risks of asthma, atopic dermatitis and chronic lymphocytic leukemia.

This is all starting to sound depressing until I dig deeper and discover that the average population's risk of getting these diseases is small in the first place and that my "above average" susceptibility is still a low risk overall. In addition, a lot of the research linking particular genes to medical conditions is preliminary and therefore not yet conclusive.

And just as I have elevated risks of some conditions, I have reduced

---

*Actually, the website tells me it's a condition where the immune system "mistakenly attacks the mucous membranes and moisture-secreting glands of the eyes and mouth." Nice.

risks of others including lung cancer and melanoma. Apparently I am also less likely to be addicted to nicotine than your average Joe, but more likely to get hooked on heroin should I give it a try. So, guess I'll give that a miss then.

I pay more attention to the risks given four stars (although these still come with various caveats). It seems my rear end is the weak point. Apparently I am 1.47 times more likely to get prostate cancer than your average bloke, and 1.43 times more likely to get colorectal cancer too. I'm not that bothered. Knowledge is power and I can now try to proactively reduce my risk of both conditions through some simple lifestyle choices. For instance, eating fish has been linked to reducing the risk of prostate cancer and I like fish, so bring on the (responsibly farmed) salmon.[2] I'll make a point of getting screened regularly to detect anything untoward early on as well.

I'm actually less struck by the information I'm reading than by how it came to me. I'm looking at genomic data (reminding me of my trip to see George Church at Harvard Medical School) that has been provided by a piece of biological nanotechnology (bringing to mind a coffee shop in Palo Alto and a talk with Eric Drexler). On the grass next to me is the bag given to me by Konarka. Its organic flexible solar panel (that rolled off that "big printer" I saw in New Bedford) has been charging the mobile phone I am now using to access the Internet (cofathered by Vint Cerf, whom I met in Washington, D.C). Thinking back to my conversation with Ray Kurzweil about exponentials, I realize that all of this technology would have cost millions of dollars ten years ago. Somewhere on my computer is Hod Lipson's "truth finding" software brain that I marveled at in Ithaca. Given the right data, my humble laptop is now capable of discovering new knowledge.

The sun pokes through the clouds and my thoughts move to what I've learned about our battle with climate change. Looking into the sky I remember Klaus Lackner's machines that can take CO_2 out of the air. I remember how that CO_2 can be used as a raw material to create liquid fuels that could replace gasoline, harnessing the power of synthetic biology. Maybe such fuels will one day power spacecraft like those I saw in Mojave? I think of Vicki Buck and her algae fixation and the Black Phantom that takes waste biomass and turns it into useful products, including carbon-sequestering charcoal. The grass between my fingers brings to mind my journey through the Australian bush with Bruce and Tony and those fence lines that separated areas of growth from bare soils, signaling a system of agriculture that is rejuvenating biodiversity and profits on Australian farms simultaneously, all the while sequestering carbon into the soil.

As I sit on this hill writing these final words, I know that what I have seen is only the tiniest fragment of the innovation going on and an even tinier fraction of the innovation yet to arrive. I know that advances are coming faster than Paralympic athlete Oscar Pistorius, whose carbon fiber blades give athletics officials the willies about his "unfair advantage." I have to make peace with the fact this book is already a historical document. It's less a posed portrait, more a blurred snapshot of a rapidly innovating world. An action shot, if you like.

In his book *Engines of Creation* nanotechnologist Eric Drexler approaches the future by asking three questions: What is possible? What is achievable? What is desirable? The first question seems easy to answer. As we learn to control the very atoms of matter, the mechanisms of biology and the power of computation, there is in fact very little that we can't do in a physical (and indeed virtual) sense. Solutions to climate change? Already developed. An end to the energy crisis? No sweat, sign on the line. Increased food production? Why didn't you ask? Holidays in space? Join our frequent flyer program. Cheaper, better manufacturing through nanotechnology? We're working on it.

But when we ask what is *achievable*, well that's a different story. Because what we achieve will largely be determined by what we collectively decide is desirable. As George Church told me all those months ago at Harvard Medical School while we discussed personal genomics, "The only thing that puts this kind of medicine far away is really will, right? The question is, how motivated are we?"

A lack of will isn't our only failing, though. Sometimes we're too eager to accept the easy promise of technology too. History is replete with techno-optimists who claimed the next revolution from science would usher in a golden age. The most bombastic of their number believe there is no problem technology cannot solve. Two years after the optical telegraph station that once stood behind me was erected the *Encyclopaedia Britannica* boldly suggested the networks of which the station was part would soon settle disputes between nations in hours instead of years. Over two hundred million people have died in conflict since.

That said, I've seen so much on my trip that has made me feel optimistic. It's hard not to smile when you see what George Church is doing to the world of medicine. He's providing tools that have the potential to save millions of lives (although as a scientific adviser to 23andMe, he's also partially responsible for the death of a few extra fish, now that I'm going to be eating more). It's difficult not to feel a glimmer of hope when you meet Klaus Lackner, or see Carbonscape's Black Phantom, or realize the potential of nanotechnology to underpin revolutions in energy production, health care and sanitation. The other day my editor asked me to list some of the things I'd discovered on my journey, to get their sales team enthused about the book. Off the top of my head I wrote a chapter-sized document broken into subsections. Here are just some of the titles I gave them:

Replacement body parts grown from patients' own stem cells are already here. Bionics could soon be able to help the "handicapped" out-perform the rest of us. The scene in *The Empire Strikes Back* where Luke Skywalker gets a new hand has already happened for real. We're learning to recode cells—curing disease from the inside. We're all living longer, and globally the gap between rich and poor, healthwise, is narrowing. The genomics revolution means lifesaving drugs we can't use today because of regulation could be released to the market—and make all drugs cheaper. We can

stop pandemics (even those made by bioterrorists). There's no shortage of "fossil fuels," we just won't make them out of fossils (and they'll
be carbon neutral). Solar power is finally coming. There's an "off-grid"
revolution starting up. We're harnessing the power of life to make everything from drugs to plastics—also known as "How bacteria are being
used to make things cheaply." Nanotechnology could change everything,
including ending capitalism, and ushering in an age of abundance. The
machines will finally become friendly (the cutting edge of robotics looks
nothing like *The Terminator*). We can take CO_2 out of the air, reduce carbon
emissions and increase food production (while reducing its cost) simultaneously. We're heading back to space. The Internet is a laboratory for
human good . . .

. . . and so on.

But there's one thing technology cannot solve. A lack of imagination.

How is it that we have not managed to equitably distribute the wealth
we create? There are millions of people who do not feel the same dividend
of increased life spans and better health care that industrial capitalism has
brought to some. One in five child deaths is the result of diarrhea.[3] The
same proportion dies of malaria. (In Africa, a child dies every forty-five
seconds from the disease.[4]) The Rwandan genocide was estimated to have
killed eight hundred thousand people in just a hundred days.[5] The pre-
dominant technologies on show there were weapons of slaughter. Those
people died in the same world where someone thinks it's a good idea to
manufacture a lady's handbag that costs a quarter of a million dollars.[6]

Should I do the whole list? The catalogue of injustice, violence, prejudice and inequality that still besets us? I could fill this book again if I did.
But as far as books go, that market is pretty much covered. In *The Rational
Optimist* Matt Ridley recalls browsing the current affairs section of an
airport bookshop and, despite volumes on offer from authors as diverse
as Noam Chomsky, Barbara Ehrenreich, Al Franken, Al Gore, John Gray,
Naomi Klein, George Monbiot and Michael Moore, not seeing a single
optimistic volume. "All argued to a greater or lesser degree that (a) the

world is a terrible place (b) it's getting worse (c) it's mostly the fault of commerce and (d) a turning point has been reached," he lamented.[7]

I think Ridley is being a bit harsh on those books. I interpret them as heartfelt calls to action: dire warnings followed by suggestions about what to do to make things better. You don't have to agree with everything or indeed anything they say, but if it is doom-mongering, it is at least *proactive* doom-mongering, motivated by the belief that things can get better or that the worst can be avoided. Concerned friends don't always tell you what you want to hear.

My good friend Eliza says, "The vast majority of the world's natural resources are controlled and divvied up by old men sending young men to war to protect their empires of traditional power, and that has not yet changed much." Eliza's great at this kind of thing. She's just the sort of person to remind me that some histories of Christopher Columbus, whose quote I use to open this final chapter, mention his habit of murdering natives (and I value her for it). She really *could* have filled this book with all the things that are, and have been, wrong with the world. She's a walking encyclopedia of doom who doesn't see the contradiction in being pessimistic and progressive. "Why can't you be pessimistic but remain committed to all progressive options?" she asks. "I have an issue with the idea that we should all be optimistic, it seems to be disrespectful to people who are suffering and dying now because of our messed-up world." She's echoing the sentiments of the French writer Romain Rolland, who wrote, "What I especially love . . . is this intimate alliance—which for me makes the true man—of pessimism of the intelligence, which penetrates every illusion, and optimism of the will."[8]

Sitting in Telegraph Hill Park, London, I've come back to the thoughts that first occurred to me sitting on the steps of the New York Public Library—that any dream of the future must be shaped and influenced by the society in which it is dreamed. I do not aspire to the same future that my cave-dwelling ancestors did. Ray Kurzweil remarked, "If you asked people ten thousand years ago what would solve the problems of the world, they'd say, 'If we can find a way to move bigger stones in front of our caves to keep the animals out and a way to keep these fires going, that would be utopia and we'd solve all our problems.'" Those living in hippie communes do not dream the same dreams as those who work on Wall Street trading floors. It's back to what John Seely Brown and I had discussed—how the frame shapes the picture more than we comprehend.

For all the marvels I've seen on my travels, I realize that what has impressed me the most is not the innovations on display but the people I have met. Because nearly all of them seem driven by one question: How can I make the world better? This question has framed their lives more than anything else. And you know what? Nearly all of the people who I feel are asking themselves that question seem really happy.

My personal manifesto for approaching the future came together on the steps of the New York Public Library and I'm feeling comfortable with it now. I don't expect anyone else to follow it, but I'm going to give it a go. I will endeavor to embrace the exponential, come to terms with the fact evolution isn't over, understand that institutions that don't innovate in the way they organize themselves (including the ones I work for) might have office space going soon, adopt a questing disposition and measure my value by what I create, not what I own.

The future is going to be a rocky ride. Many new technologies will have their Three Mile Islands and be coopted in the service of more 9/11s. People will still fight. Injustices will not disappear. Andrew Lloyd Webber musicals will still be staged. But now that my journey is complete, I steadfastly refuse to believe that human society can't grow, improve and learn; that it can't embrace change and remake the world better, that we can't make the underlying trend upward. The musician Nick Cave summed it up, I think: "People think I'm a miserable sod but it's only because I get asked such bloody miserable questions," he said.

Here's a question that is neither miserable, nor cheerful. It's the one I started out with.

"What's next?"

I asked it because I wondered what my life might look like. What I've discovered is that my question is its own answer. There is a new story of humanity waiting to be told if we respond well to its challenge. I reckon we can have a go, don't you think?

Right, I'm off to eat some fish.

ACKNOWLEDGMENTS

When I started writing this book I had no idea what I was getting myself into. With such a broad scope it would have been impossible to complete without the help of many, many people who generously shared their time and knowledge. "Acknowledgments" doesn't seem the right title for this. I'd like to do more than "acknowledge" these people—I'd like to take them all down to the pub, buy them a drink and tell anyone who will listen how brilliant they all are.

First and foremost, I must thank Caroline. She'd hate it if I listed all the reasons why here, so I wont.

There are some other women in my life—notably, my colleagues at Flow Associates. Bridget McKenzie let me run off into the wilds while she held the fort, doing incredible work on the interplay of learning, culture, ecology and technology and still finding time to discuss the issues in the book, raise important challenges, be one of my best friends and, along with the poker-playing demon that is her husband, Brian, welcome me into her home with offers of wine and food. The other two Flow-ers, Katherine Rose and Eliza Hilton, have been amazing friends and commented with sagelike wisdom on the manuscript, too, all while helping to set up our office in India. On matters of manuscript reading, I am also deeply indebted to my wonderful niece, Amy Stevenson, her fiancé, Mitchell Burns, and the incomparable Sandra Higgison for taking the time to read many chapters and comment on them. Businesswise, apologies must also go to Quentin Cooper and Elaine Snell, my codirectors of live science agency ReAgency, for not being around to do all the things I said I would.

A special mention must go to my agent Charlie Viney for his razor-sharp criticism, faith from the beginning, putting up with my odd turns of phrase, and, with his amazing partner, Sally, and family, bringing me into his house for sparkling conversation and general encouragement. A prince among men.

Charlie, by coincidence, lives on the same street as Philip Groom, a man who needs heralding for his intelligence, grace and Picasso-esque insight into matters of growth and engine maintenance. Without his good counsel I'm not at all convinced this book would have ever made it off the starting blocks.

I've also been blessed with a great team at my publishers Penguin/Avery. I'd like to thank Rachel Holtzman, my editor, for her light but astute editing of the manuscript and being incredibly easy to work with; Travers Johnson for clearance heroics and general easygoingness (Your Time Is Gonna Come); Beth Parker, Jessica Chun and Lisa Johnson for spreading the word; William Shinker for saying "yes;" Dorian Hastings, Jill Newman, and Anna Jardine for their diligent copy-editing (in return I give you the word 'boffin'); and Nellys Li for the great cover design. I owe you all a round of "Optimists" next time I'm in NYC.

This book relies heavily on the wisdom and hard work of all the scientists, engineers, thinkers, farmers, writers and other superbrains that gave me their time. I've had some of the most entertaining (and mind-blowing) conversations and experiences of my life on this journey. It is thanks to these people that I was able to glimpse something of our future. All of them (with a couple of nameless exceptions) also found the time to read over my words and offer comment, corrections or useful suggestions. Any remaining errors pertaining to their work are mine alone. In rough order of appearance, I am delighted to thank the following list of extraordinary people:

- Nick Bostrom of the Future of Humanity Institute at Oxford University really did blow my mind and helped me to dive in at the deep end.
- George Church of Harvard Medical School shared an inordinate amount of his time, given how busy he is redefining the worlds of medicine and biology. Someone give that man a Nobel Prize
- Mark Bedau, Professor of Philosophy and Humanities, Reed College. Sadly a lot of my meeting with Mark didn't make it into the book (although the bits that do I think are some of the most important). As one of my early interviewees, he calmly dealt with some of my more naive questions and did it all in front of the camera crew who were following me at the time.
- Cynthia Breazeal. Cynthia tends not to give face-to-face interviews so I'm particularly grateful to her for agreeing to meet and introducing me to her sociable

robots and the ideas they unleash. I still think her lab needs a tidy-up, though. Do a search for "Leo Robot" on YouTube and be amazed. (Thanks, too, to Dan Stiehl for showing me the cuddly machines.) While I was at the Media Lab, I also spent a fascinating hour and half with Bill Mitchell of the Smart Cities group, a true outback kid now heading up one of the most interesting labs in America. Thanks, Bill.

- Hod Lipson couldn't have been friendlier, and took it upon himself to keep me up-to-date with his work long after I'd been to see him at Cornell University. His curiosity about curiosity is invigorating.

- Eric Drexler, like Cynthia Breazeal, gave me a rare face-to-face interview. Beyond that, he took extreme care to explain many subtleties concerning nanotechnology in long e-mail exchanges following our meeting. Head on over to e-drexler.com for an entertaining read, including the online version of *Engines of Creation: The Coming Era of Nanotechnology* (which Eric kindly gave me permission to quote from liberally).

- Dr. Richard Jones. I didn't get a chance to meet the professor of physics at Sheffield University in person, but over the phone and in e-mails he provided clear insights into the profession of nanotechnology and the schisms therein. His blog www.softmachines.org is well worth a visit.

- Stuart Witt, Dick Rutan and Jeff Greason, whom I met at Mojave Space Port, were everything a writer could wish for, full of adventure and strong opinions, and up for an Elk Burger. Thanks also to Xavier Claramunt and Marc Zaballa-Camprubí of Galactic Suite, Barcelona, for gin and tonics and the three questions that matter (apologies that more of our chat didn't make the final cut).

- Vint Cerf must be thanked first and foremost for cofathering the Internet (I couldn't have written this book without it), but also for being an all-round good sport and a generous interviewee and for the nice merlot we had after our discussion (Vint collects wine, "but only to drink" he assured me). Also from Google I must thank Dan Reicher for talking energy efficiency, and Peter Norvig, head of research, for taking an hour out to discuss a range of topics (our conversation didn't make it into the book but I enjoyed it greatly, nonetheless).

- Klaus Lackner and Wally Broecker. Getting an interview with either of these guys is tricky—getting them both at the same time is like a geek-jackpot. (I have to thank my UK editor Mark Ellingham at Profile Books for making this possible.) Klaus may just help save the planet. Thank you, Klaus. Wally is one of the people who made us aware we need to. Thank you, Wally.

- The good people of Konarka. Tracy Wemett helped me appreciate life (and not just because of the near-death experiences), became a friend in Boston and is a beacon of general all-round good-eggness. Thanks for everything, especially the Lucky Luckys. In addition I'd like to thank Konarka CEO Rick Hess for taking me off-grid; Larry Weldon, vice president of manufacturing, for showing me his butt splicer; cofounder Howard Berke for being nicely controversial (he didn't make it into the book, either); and Terri Jordan and Gilles Dennler for fact-checking.

- Vicki Buck is without doubt one of the most generous people on the earth. She spent a week introducing me to a whole host of fascinating people in New Zealand, including flying me to Auckland to meet Rob Fenwick, Vincent Heeringa and Sam Tobin, none of whom made it into the manuscript but all of whom informed it (I am particularly indebted to Sam for a rewarding e-mail correspondence). It was through Vicki that I also met Nick Gerritsen, the firebrand knowledge broker (who became a generous correspondent, firing off bits of research or thoughts on the future at regular intervals) and Tim Langley, chairman of Carbonscape (who gave me one of the most treasured moments of my journey on his boat *Faith,* and somehow calmed my soul).

- Australian double act Tony Lovell and Bruce Ward took an entire week out of their schedules to show me a farming revolution and introduce me to sheep metrics. We had a ball and are now trading insults like old friends. Old Love-with-two-L's says he's got a few new angles on Pom bashing he'd like to try out on his next trip to London, and I'm ready for them. It was thanks to Bruce and Tony that I got to meet the farmers Tim and Karen Wright, Andrew and Megan Mosely, Grahame Strong, Ken Haylan and Michael Coughlan—a coterie of agricultural geniuses who kindly shared their stories with me.

- President Mohamed Nasheed. Presidents are busy people, especially when they're the first democratically elected leader of a nation. Meeting Nasheed was a highlight, showing that almost anything is possible, even when the hour is at its darkest. We could do with a few more world leaders like that. (A mention must also go to Mohamed Ziyad, executive services secretary, who, after initially telling me there was "no chance" of my getting an interview with the president, helped to make it happen).

- Ray Kurzweil gave me more time than a man as busy as he would reasonably want to, and got me into exponential trousers. You can't say that about many people. If you fancy delving deeper into Ray's worldview, then *The Singularity Is Near* will keep you going for a week or three.

- Juan Enriquez was a model of generosity, both intellectually and in introducing me to the good people of TED. As evolved human beings go, Juan is one of my favorites. His book *As the Future Catches You* is highly recommended.

- John Seely Brown (on the recommendation of Vint Cerf) saw me at impossibly short notice. He talked about cultivating serendipity, and this meeting, for me, was an example of it. His insights helped me pull many strands of thought together. Do read his (coauthored) latest, *The Power of Pull*.
- Chris Anderson should be thanked for taking the time to see me in between all the other things he does, but more widely for doing the TED thing. If you haven't seen TED.com then I recommend you get over there and prepare to get your frames rejigged. As my final port of call, he couldn't have made a better interviewee.

Beyond my interviewees, thanks go also to Rodney Brooks and Stewart Brand for allowing me to quote from their work. Do read *Robot: The Future of Flesh and Machines* and *Whole Earth Discipline*. I enjoyed them both.

One thing I realized very soon into my trip is that most organizations only survive because of their administrators. I owe a huge debt of gratitude to everyone who helped me get in contact with my interviewees and secure times to see them. I tell you, MIT would fall over if it wasn't for people like the formidable Polly Guggenheim (Personal Robotics Lab) and Cynthia Wilkes (Smart Cities Group). Warm appreciation also flows to Yveta Marsarova (Harvard Medical School) and Gabriel Terrizzi (Cornell). Carrie Howell (and a volcano) arranged my meeting with John Seely Brown, while Sarah Reed of Kurzweil Technologies persevered with her boss and finally secured me a meeting with Ray Kurzweil himself. Lisa Goldberg helped set up my time with Juan Enriquez and assisted in tracking down some key e-mail addresses too.

Public relations professionals are often much maligned, but a good one is like a draft of fresh water to a writer juggling travel timetables, personal deadlines and the needs of research. Thanks then to Felicia Spagnoli at Joule Biotechnologies for putting me in touch with John Ward (Joule's senior vice president of production) and, in doing so, confirming that the idea for a carbon neutral petrol station that pulls its raw materials out of the sky isn't bonkers. Paul Roberts got me to an underwater cabinet meeting in the Maldives, and for that I will always be grateful. Laura Galloway of Galloway PR organized my meeting with Chris Anderson at TED, helped with clearances to quote numerous TED talks, took me out to dinner and remains generous, helpful and encouraging. She is also owed a cocktail the next time I'm in NYC. Rachel Nagler at Rubenstein PR organized for me to have my genetic profile processed by 23andMe. Rachel Whetstone, Anthony House and Carla LaFever (all of Google) formed a procession that led to Vint Cerf, and Nikki Fenwick and Aya Okuma helped me hook up with Dan Reicher.

I've made a brace of new friends on my travels (and reconnected with some old ones). Kind-hearted individuals helped me find accommodation, and food and water and/or welcomed me into their homes. In Christchurch, Jane Ellis invited me over for family dinner. In Auckland, Anna and Robert Booy kept me entertained at high altitude. In New York, Adrian Mukasa provided hilarious company and found me an apartment (now get behind a mic!). In California, Karen Carney and boyfriend Bruce provided me with lodging and fine company. So, too, did Arthur Gleckner and Kristine Kelly. I bumped into Arthur and Kristine in Boston, then in Sydney and finally, when they realized I wasn't stalking them internationally, they generously offered me their spare room while I grappled (intellectually, you understand) with Eric Drexler. Arthur, who works at Google, also helped smooth my time at the Googleplex. Sonu and Eva Shivdasani generously had me as their guest at the Soneva Fushi resort in the Maldives, and while I was there, Anke Hofmeister, Sophy Williams, Hannan and Alan Thomas and his wife made me feel incredibly welcome. (The stargazing with Alan was one of the highlights of my journey). During my time at Soneva Fushi, Harald Willenbrock and Peter Rigaud made just the right amount of mischief. Thanks also to Per Jensnaes, head of the International Federation of Red Cross and Red Crescent Societies in the Maldives, who entertained me for dinner in the capital, Malé. In Sydney, Renate Ruge helped me find calm, and Christine Axton laid some ghosts to rest while still shining like the brightest star. A special mention, too, must go to my friend the neuroscientist Colin O'Carroll for Vegas, NYC and all the chat. You're a hero.

An army of others helped me out with bits of research, including Malcolm Bacchus, who told me all about the optical telegraph station in my local park (and bought me a beer while he was doing it), Phillip B. Stark of the Department of Statistics at the University of California, who gave me the lowdown on Internet porn statistics, and Dr. Anthony McCann of the University of Ulster, who provided some nice pointers on war, peace and interconnectedness. Jim Aach was incredibly generous in discussing nuclear power (and I'm sorry this didn't make it into the book, Jim!). Jim's novel *Rad Decision* is a gripping thriller set in the nuclear industry. Similarly not in the book are Stephen Aguilar-Milan (who was the first person to agree to meet me and got me all warmed up and ready for the future) and Tim Tully of Dart Neuroscience (thanks for the story about Pavlov's dogs, Tim).

Heartfelt gratitude goes to Dr. Emmanuel Skordalakes, assistant professor of gene expression and regulation at the Wistar Institute (and, coincidentally, my landlord) for reviewing the chapters relating to synthetic biology and genetics, as well as funding a new kitchen. I must also mention the brilliant writers Robert Kunzig, Bob

Henson and Chris Goodall for looking over the chapters on climate change and solar power. Robert's book, written with Wally Broecker. *Fixing Climate* is a must-read history of climate science, Bob Henson's *Rough Guide to Climate Change* is truly excellent, and Chris Goodall's *Ten Technologies to Save the Planet* should be read by, well, almost everyone. Martin Birchall, professor of laryngology, also checked over the section on Claudia Castillo and deserves thanks not only for that but also pioneering new stem cell therapies.

I squeezed in a few stand-up gigs while I was traveling and would like to especially thank Jon at Mottley's Comedy, Boston; Rick Jenkins of The Comedy Studio, Harvard; and Darren Sanders at The Laugh Garage in Sydney for putting the foreigner on the bill and for being oases of company and tequila on a long journey.

For a large part of the time I was writing this book. I had a broken wrist (which, as you can imagine, isn't the best thing for a writer). However, thanks to Dr. Mark Philips and his colleagues at Kings College Hospital, London, I got fixed without the need for surgery. Mark, time for that pint I owe you.

Thanks also to Peter Schöni for the genius mixology of the "Optimist" cocktail; the brilliantly named Susannah and Ya'Acov Darling Khan for saving the footage; Greg Atkins, Tom Keeling and Natalie Spanier for faith and support (it will happen!); Meredith McDonough for being a George; Tony Kensington for the his lovely figure(s); Lucy, Andrew and Abigail Billen for ending chapter one; my family for the genome and never letting me get above myself; Imran Yasin for living an inspirational life; Hannah Williams for believing and persevering (it meant a lot and, not long now); and René Hen, director of the division of integrative neuroscience at Columbia, for the story about the mushrooms.

Final mentions must go to Taragh Bissett, Nikki Roswell, Paul Milnes, Holger Emrich, Duncan Brown, Rebecca Suffling and Anouchka Grose for friendship when it mattered; Kat Arney for helping me find the Golem; Jon Collins for the progressive advice; the inspirational Andy Ross for insight and encouragement; and finally, special comedy-tinged theatrical thanks to Jack Milner for putting out the chairs and getting me rolling. Jack, you rock.

If I missed anyone (and I'm sure I have), then apologies, but it was a long trip and I met a lot of people.

My round.

NOTES

Chapter 1. "The World's Most Dangerous Idea"

1. "Birthday Skydive for 90-year-old," BBC News, 8 Aug. 2006. http://news.bbc.co.uk/1/hi/england/5257354.stm.
2. Office for National Statistics, "Social Trends No.40 – 2010 edition," 6 July 2010. Data set to support Figure 7.1: 'Expectation of life at birth: by sex' - table 07_01.xls http://www.statistics.gov.uk/downloads/theme_social/Social-Trends40/Social-Trends-40.zip). Full report: http://www.statistics.gov.uk/downloads/theme_social/Social-Trends40/ST40_2010_FINAL.pdf.
3. Elizabeth Arias, "National Vital Statistics Reports, Volume 58, Number 21," Division of Vital Statistics, National Center for Health Statistics. http://www.cdc.gov/nchs/data/nvsr/nvsr58/nvsr58_21.pdf.
4. Office for National Statistics, "United Kingdom Interim Life Tables, 180-82 to 2006-08" http://www.statistics.gov.uk/downloads/theme_population/Interim_Life/ILTUK0608Reg.xls.
5. U.S. Social Security Administration, "Period Life Table 2006". http://www.ssa.gov/OACT/STATS/table4c6.html.
6. "Life Expectancy to Soar," BBC News, 9 May 2002. http://news.bbc.co.uk/1/hi/health/1977733.stm.
7. S. Gottlieb, "Increases in Life Expectancy Likely to Be Smaller in the Future," *British Medical Journal* (322):512. http://www.bmj.com/cgi/content/full/322/7285/512.
8. Francis Fukuyama, "Transhumanism," *Foreign Policy*, Sept./Oct. 2004.
9. Julian Huxley, *Religion without Revelation* (Greenwood Press, reprint, 1979).
10. Dr. Aubrey de Grey, "We Will Be Able to Live to 1,000," BBC News, 3 Dec. 2004. http://news.bbc.co.uk/1/hi/uk/4003063.stm.
11. "Aubrey de Grey says we can avoid aging," TEDGlobal website, 2005. Full talk available at http://www.ted.com/talks/aubrey_de_grey_says_we_can_avoid_aging.html.
12. "Nick Bostrom on our biggest problems," TEDGlobal website, 2005. Full talk available at http://www.ted.com/talks/nick_bostrom_on_our_biggest_problems.html.

13. Internet Encyclopedia of Philosophy, "Ge Hong." http://www.iep.utm.edu/g/gehong.htm.

14. Jason Pontin, "Against Transcendence—When technology appropriates the transcendental it becomes science fiction," *MIT Technology Review*, February 2005. http://www.technologyreview.com/communications/14146/?a=f.

15. Jason Pontin, "Is Defeating Aging Only a Dream?" *MIT Technology Review*, July 2006. http://www.technologyreview.com/sens/.

16. Harry Collins and Trevor Pinch, *The Golem: What You Should Know About Science* (Cambridge, UK: Cambridge University Press, 1993), "Ch. 1: Edible Knowledge, the Chemical Transfer of Memory."

17. www.nickbostrom.com.

18. Maria Blasco et al, "Telomerase Reverse Transcriptase Delays Aging in Cancer-Resistant Mice," *Cell* 135(4):609–22. http://www.cell.com/abstract/S0092-8674%2808%2901191-4.

19. The Nobel Prize in Physiology or Medicine 2009, "For the discovery of how chromosomes are protected by telomeres and the enzyme telomerase," Nobel Prize.org. http://nobelprize.org/nobel_prizes/medicine/laureates/2009/.

20. D. Ornish et al., "Increased Telomerase Activity and Comprehensive Lifestyle Changes: A Pilot Study," *The Lancet Oncology* 9(12):1124. http://www.ncbi.nlm.nih.gov/pubmed/18799354; and also Christian Werner, M.D., et al. "Physical Exercise Prevents Cellular Senescence in Circulating Leukocytes and in the Vessel Wall, *Circulation* (20):2438–47, © 2009 American Heart Association, Inc. http://circ.ahajournals.org/cgi/content/short/120/24/2438.

21. Sarah Boseley, "'My unique chance': How Claudia Castillo's Landmark Operation Changed Her Life," *Guardian*, 19 Nov. 2008. http://www.guardian.co.uk/society/2008/nov/19/organ-donation-medicalresearch.

22. "Stem Cell Facts. The Next Frontier?" International Society for Stem Cell Research. http://www.isscr.org/public/ISSCR08_PubEdBroch.pdf.

23. "Stem Cells Made to Mimic Disease," BBC News, 7 April 2008. http://news.bbc.co.uk/1/hi/health/7334365.stm.

24. Krista Conger, "Dramatic Transformation: Researchers Directly Turn Mouse Skin Cells into Neurons, Skipping IPS Stage," Stanford School of Medicine website, 27 Jan. 2010. http://med.stanford.edu/ism/2010/january/wernig.html.

25. Yuin-Han Loh et al, "Reprogramming of T Cells from Human Peripheral Blood," *Cell: Stem Cell*, 2 July 2010, volume 7, issue 1. http://www.cell.com/cell-stem-cell/fulltext/S1934-5909(10)00286-9.

26. "U.S. Doctors Use Stem Cells to Replace Organs," *Stem Cell News*, April 2006. http://www.stemcellnews.com/articles/stem-cells-to-replace-organs.htm.

27. Wake Forest Institute for Regenerative Medicine. http://www.wfubmc.edu/WFIRM/.

28. Deane Morrison, "Researchers Create a New Heart in the Lab, *UM News*, 14 Jan. 2008. http://www1.umn.edu/news/features/2008f/UR_167337_REGION1.html.

29. "University of Minnesota Finalizes License Agreement to Form Start-up Company Based on Doris Taylor Research," University of Minnesota press release, 16 Feb. 2010. http://www1.umn.edu/news/features/2010/UR_CONTENT_179497.html.

30. Ray Kurzweil, *The Singularity Is Near: When Humans Transcend Biology* (New York: Viking Press, 2005).

31. George J. Annas, Lori B. Andrews, Rosario M. Isasi, "Protecting the Endangered Human: Toward an International Treaty Prohibiting Cloning and Inheritable Alterations," *American Journal of Law and Medicine* 28(2 & 3):151–78.

32. Gapminder.org statistics website: http://www.gapminder.org/?s=life+expectancy.

33. Paolo Rama, M.D., et al, "Limbal Stem-Cell Therapy and Long-Term Corneal Regeneration," *The New England Journal of Medicine*, 8 July 2010.

34. Julian Smith, "We Have the Technology," *New Scientist*, 3 Jan. 2009.

35. "Interview with Dr. Sarah Rankin," *Today*, BBC Radio 4, 9 Jan. 2009.

36. Andy Coghlan, "Heal Yourself, With Stem Cells' Help," *New Scientist*, 17 Jan. 2009.

37. Juan Enriquez, "Homo Evolutis," in *What are you Optimistic About?* (London: Simon & Schuster). Also available online at http://www.edge.org/q2009/q09_1.html#enriquez.

38. Sander Olson, "Interview with Aubrey de Grey," Next Big Future, 17 July 2009. http://nextbigfuture.com/2009/07/aubrey-de-grey-interview.html.

Chapter 2. "The Most Wondrous Map"

1. Juan Enriquez, "Private vs. Public Sequencing Effort," in *As The Future Catches You* (New York: Random House/Three Rivers Press, 1st pbk. ed.), p. 110.

2. James Shreeve, "Craig Venter's Epic Voyage to Redefine the Origin of the Species," *Wired*, Aug. 2004. http://www.wired.com/wired/archive/12.08/venter.html.

3. Thomas Goetz, "How the Personal Genome Project Could Unlock the Mysteries of Life," *Wired*, Aug. 2008. http://www.wired.com/medtech/stemcells/magazine/16-08/ff_church?currentPage=2.

4. David Concar, "Ecstasy Has Dramatic Effect on Parkinson's Symptoms," *New Scientist*, Nov. 2002. http://www.newscientist.com/article/dn3015-ecstasy-has-dramatic-effect-on-parkinsons-symptoms.html.

5. J. H. Thomas, "Genetic Analysis of Defecation in Caenorhabditis elegans," *Genetics* 124(4):855–72, © Genetics Society of America. www.genetics.org/cgi/reprint/124/4/855.pdf.

6. "Frequently Asked Questions," Archon Genomics Xprize website, 12 Jan. 2010. http://genomics.xprize.org/archon-x-prize-for-genomics/frequently-asked-questions.

7. Ayala Ochert, "Deconstructing DNA," *New Scientist*, May 1998. http://www.newscientist.com/article/mg15821345.000-deconstructing-dna.html?page=1.

8. "Map of Elephant DNA Reveals Trail of Ivory Smugglers," *South African Mail & Guardian*, 28 June 2009. http://www.mg.co.za/article/2009-06-28-map-of-elephant-dna-reveals-trail-ivory-smugglers.

9. E. D. Pleasance et al., "A Small-cell Lung Cancer Genome with Complex Signatures of Tobacco Exposure" *Nature* 463(14 Jan. 2010):184-90. http://www.nature.com/nature/journal/vaop/ncurrent/full/nature08629.html; and also E. D. Pleasance, R. K. Cheetham et al., "A Comprehensive Catalogue of Somatic Mutations from a Human Cancer Genome," *Nature* 463(16 Dec. 2009):191–96. http://www.nature.com/nature/journal/vaop/ncurrent/full/nature08658.html Summary information: "Lung Cancer and Melanoma Laid Bare," Wellcome Trust/Sanger Institute press release, 16 Dec. 2009. http://www.sanger.ac.uk/about/press/2009/091216.html.

10. B. J. Morrison McKay and Clare Sansom, "Webb Miller and Trey Ideker to Receive Top International Bioinformatics Awards for 2009 from the International Society for Computational Biology," *PLoS Computational Biology*, 24 April 2009. http://pubmedcentralcanada.ca/articlerender.cgi?artid=135702.

11. George Church, "Participant #1," Personal Genome Project. http://www.personalgenomes.org/public/1.html.

12. World Health Organization International Agency for Research on Cancer, *IARC Monographs on the Evaluation of Carcinogenic Risks to Humans: Tobacco Smoke and Involuntary Smoking* (Lyon,

France: IARCPress, vol. 83, 2004). http://monographs.iarc.fr/ENG/Monographs/vol83/volume83.pdf.

13. Natalie Angier, *The Canon: A Whirligig Tour of the Beautiful Basics of Science* (London, UK: Faber and Faber, 2008, pbk. ed.), "Ch. 7: Molecular Biology: Cells and Whistles," p. 211.

14. "We Have Learned Nothing from the Genome," *Der Spiegel*, 29 July 2010. http://www.spiegel.de/international/world/0,1518,709174,00.html.

15. Christopher P. Adams and Van V. Brantner "Estimating the Cost of New Drug Development: Is It Really $802 Million?" *Health Affairs* 25(2):420-28. http://content.healthaffairs.org/cgi/content/abstract/25/2/420.

16. Emily Singer, "Complete Genomics Releases $5000 Genome Data," *Technology Review* blog, 6 Feb. 2009. http://www.technologyreview.com/blog/editors/22618/.

17. Emily Singer, "The $100 Genome," *Technology Review*, 17 April 2008. http://www.technologyreview.com/Biotech/20640/.

18. Emily Singer, "The $30 Genome?" *Technology Review*, 7 June 2010. http://www.technologyreview.com/biomedicine/25481/.

19. "The Promise of Personalized Medicine," Archon X Prize for Genomics website. http://genomics.xprize.org/archon-x-prize-for-genomics/the-promise-of-personalized-medicine.

Appendix: Damned Clever, Those Russians

1. Gil Atzmon et al., "Genetic Variation in Human Telomerase Is Associated with Telomere Length in Ashkenazi Centenarians," in *Proceedings of the National Academy of Sciences of the United States of America*, 13 Nov. 2009. http://www.pnas.org/content/early/2009/11/11/0906191106.abstract.

2. Robert Roy Britt, "One Key Found for Living to 100," Live Science website, 12 Nov. 2009. http://www.livescience.com/health/091112-anti-aging-telomeres.html.

3. Overview of BioTime's Reversal of the Aging of Human Cells," Biotime Inc. website. http://www.biotimeinc.com/overview3-10.htm. Full paper: H vaziri et al, "Spontaneous reversal of the developmental aging of normal human cells following transcriptional reprogramming," Regenerative Medicine, May 2010, Volume 5. http://www.futuremedicine.com/doi/abs/10.2217/rme.10.21.

Chapter 3. "My Lab Should Not Be Trusted."

1. Ewen Callaway, "Immaculate Creation: Birth of the First Synthetic Cell," *New Scientist*, 20 May 2010. http://www.newscientist.com/article/dn18942-immaculate-creation-birth-of-the-first-synthetic-cell.html.

2. George Church, "Life After the Synthetic Cell—Now Let's Lower Costs," *Nature*, 27 May 2010. http://www.nature.com/nature/journal/v465/n7297/pdf/465422a.pdf.

3. Katie Howell, "*ExxonMobil Bets $600 Million on Algae*," *Scientific American*, 14 July 2009. http://www.scientificamerican.com/article.cfm?id=biofuels-algae-exxon-venter.

4. "LS9 UltraClean™ Fuels," LS9 website. http://www.ls9.com/products/.

5. Jennifer Kho, "*LS9 to Start Building Demo Plant, Raising $65M,*" Earth2Tech website, 25 Feb. 2009. http://earth2tech.com/2009/02/25/ls9-to-start-building-demo-plant-raise-65m/.

6. Fareed Zakaria, "A Bug to Save the Planet," *Newsweek*, 16 June 2008. http://www.newsweek.com/id/140066.

7. "Your Plastic Pal," *Economist*, 26 Nov. 2009. http://www.economist.com/sciencetechnology/displaystory.cfm?story_id=14960045.

8. David Lubertozzi and Jay D. Keasling, "Expression of a Synthetic *Artemesia annua* Amorphadiene Synthase in *Aspergillus nidulans* Yields Altered Product Distribution," *Journal of Industrial Microbiology and Biotechnology* 35(10):1191–98. http://www.springerlink.com/content/h0461211537185u0/.

9. "From SemBiosys, A New Kind of Insulin," *Business Week*, 13 Aug. 2007. http://www.businessweek.com/magazine/content/07_33/b4046083.htm. And also J. G. Boothe et al., "Analytical Characterization, Safety and Clinical Bioequivalence of Recombinant Human Insulin from Transgenic Plants," *Symbiosis*. http://www.sembiosys.com/docs/Boothe_abstract_ADA_2009_final .pdf.

10. Sheryl Gay Stolberg, "The Biotech Death of Jesse Gelsinger," *New York Times*, 28 Nov. 1999. http://www.nytimes.com/1999/11/28/magazine/the-biotech-death-of-jesse-gelsinger .html?pagewanted=1.

11. "The Case of Jesse Gelsinger," University of Pennsylvania High School Bioethics website. http://www.highschoolbioethics.org/briefs/gelsinger.asp.

12. "Guidance for Industry: Gene Therapy Clinical Trials—Observing Subjects for Delayed Adverse Events" (Rockville, Md.: U.S. Department of Healthand Human Services [FDA] Center for Biologics Evaluation and Research, Nov. 2006). http://www.fda.gov/downloads/BiologicsBloodVaccines/GuidanceComplianceRegulatoryInformation/Guidances/Cellularand GeneTherapy/ucm078719.pdf.

13. Albert Maguire et al., "Age-dependent Effects of RPE65 Gene Therapy for Leber's Congenital Amaurosis: A Phase 1 Dose-escalation Trial," *Lancet* 374(9701):1597–1605.

14. Kayt Sukel, "Gene Therapy Offers Hope for Rare Retinal Condition," Dana Foundation website, 7 Dec. 2009. http://www.dana.org/news/features/detail.aspx?id=24110.

15. H. H. Wang et al., "Programming Cells by Multiplex Genome Engineering and Accelerated Evolution," *Nature* 460(26 July 2009):894–98. http://arep.med.harvard.edu/pdf/Wang.Isaacs .Nature09.pdf.

16. Emily Singer, "A Machine That Speeds Up Evolution," *Technology Review*, 17 March 2009. http://www.technologyreview.com/biomedicine/22299/.

17. Brandon Keim, "Genome Engineering Goes High Speed," *Wired*, 27 July 2009. http://www .wired.com/wiredscience/2009/07/cellfactories/.

18. Edge.org Master Class: http://www.edge.org/3rd_culture/church_venter09/church_venter09 _index.html.

19. Rachel Nowak, "Disaster in the Making," *New Scientist*, 13 Jan. 2001. http://www.newscientist .com/article/mg16922730.300-disaster-in-the-making.html.

20. Ali Nouri, "Biotechnology and Biosecurity," lecture presented to Future of Humanity Institute, Oxford, UK, 17–20 July, 2008. http://www.global-catastrophic-risks.com/videos.html.

21. Nick Bostrom and Milan Ćirković, eds., *Global Catastrophic Risks* (Oxford, UK: Oxford University Press, 2008), p. 23. http://www.global-catastrophic-risks.com/docs/global-catastrophic-risks .pdf.

22. Stephen Herrera, "Preparing the World for Synthetic Biology," *Technology Review*, January 2005. http://www.technologyreview.com/read_article.aspx?id=14077.

23. George Church, "Let Us Go Forth and Safely Multiply," *Nature* 438(24 November 2005):423.

24. "DNA Firms Step Up Security over Bioterrorism Threat," *New Scientist*, 14 Sept. 2008. http://www.newscientist.com/article/mg19926733.500-dna-firms-step-up-security-over-bioterrorism-threat.html.

25. Michael Brooks, "The End of the World Is Not Nigh," *New Scientist*, 23 July, 2008. http://www .newscientist.com/article/mg19926663.900-the-end-of-the-world-is-not-nigh.html?page=1.

26. Dr. Iain Anderson, CBE "Foot and Mouth Disease 2007: A Review and Lessons Learned," presented to the Prime Minister and the Secretary of State for Environment, Food and Rural Affairs, London, UK, 8 March 2008. http://archive.cabinetoffice.gov.uk/fmdreview/documents/ fmd_2007_review_full.pdf.

27. "Annex Table 2 - Deaths by cause, sex and mortality stratum in WHO regions, estimates for 2002," *World Health Organization 2004 Report* (Geneva: UN World Health Organization). http://www.who.int/whr/2004/annex/topic/en/annex_2_en.pdf.

28. "Tables on the Clinical Trials of Pandemic Influenza Prototype Vaccines," World Health Organization website. http://www.who.int/vaccine_research/diseases/influenza/flu_trials_tables /en/index.html.

29. Guolin Xu et al., "A self-contained all-in-one cartridge for sample preparation and real time PCR in rapid influenza diagnosis, " *Lab on a Chip*, 2010.

30. "1918 Influenza A (H1N1) Fact Sheet," Federation of American Scientists website. http:// www.fas.org/programs/ssp/bio/factsheets/H1N1factsheet.html.

31. George Church, "Personal Genomics Will Arrive This Year, and With It a Revolutionary Wave of Volunteerism and Self-Knowledge," Edge World Question Center website, 2007. http:// www.edge.org/q2007/q07_7.html#church.

Chapter 4. Can You Turn All the Buttons On?

1. Hans Moravec, "Rise of the Robots," *Scientific American*, 23 March 2009. http://www.scientific american.com/article.cfm?id=rise-of-the-robots.

2. "National Institute of Information and Communications Technology, JVC Develop 8-mega-pixel High-resolution Robotics Technology," JCN Network website. http://www.japancorp .net/Article.Asp?Art_ID=11818.

3. Robin Marantz Henig, "The Real Transformers," *New York Times Magazine*, 29 July 2007. http://www.nytimes.com/2007/07/29/magazine/29robots-t.html.

4. Nic Fleming, "Why AI Is a Dangerous Dream," *New Scientist*, 1 Sept. 2009. http://www .newscientist.com/article/mg20327231.100-noel-sharkey-ai-is-a-dangerous-dream.html.

5. "Research Groups and Projects," MIT Media Lab. http://www.media.mit.edu/research/ groups-projects.

6. Edsger W. Dijkstra, "The Threats to Computing Science," lecture presented to the ACM 1984 South Central Regional Conference, Nov. 16-18, Austin, Texas. http://userweb.cs.utexas .edu/~EWD/transcriptions/EWD08xx/EWD898.html.

7. Rodney Brooks, "Speaker Quotes," presented to The Singularity Summit, AI and the Future of Humanity, San Francisco, Calif., Sept. 8–9, 2007. http://singinst.org/summit2007/quotes/ rodneybrooks/.

8. Steven Pinker, *The Language Instinct: The New Science of Language and Mind* (London, UK: Penguin Books, new ed., 2003).

9. Rodney Brooks, "Intelligence Without Reason," prepared for Computers and Thought, Inter-nation Joint Conference on Artificial Intelligence, April 1991. http://people.csail.mit.edu/ brooks/papers/AIM-1293.pdf.

10. Alan Turing, "Computing Machinery and Intelligence," *Mind* 59:433-60. http://loebner.net/ Prizef/TuringArticle.html.

11. Ray Kurzweil, *The Singularity Is Near: When Humans Transcend Biology* (New York: Viking, 2005), p. 92.

12. "Beyond the Doubting of a Shadow: A Reply to Commentaries on Shadows of the Mind," *Psyche* 2(23). http://psyche.cs.monash.edu.au/v2/psyche-2-23-penrose.html.

13. Douglas Hofstadter, *Gödel, Escher and Bach: An Eternal Golden Braid* (New York: Basic Books, 1979).

14. Moravec, "Rise of the Robots."

15. "Henry Markram Builds a Brain in a Supercomputer," TEDGlobal website, 2009. Full talk available at http://www.ted.com/talks/henry_markram_supercomputing_the_brain_s_secrets .html.

16. Surfdaddy Orca, "Blue Brain in a Virtual Body," *H+ Magazine*, 12 May 2009. http://www .hplusmagazine.com/articles/ai/blue-brain-virtual-body.

Chapter 5. "Like Explaining Shakespeare to a Dog"

1. Rodney Brooks, "Intelligence Without Reason," prepared for Computers and Thought, International Joint Conference on Artificial Intelligence, April 1991. http://people.csail.mit.edu/ brooks/papers/AIM-1293.pdf.

2. "When Do Humans Develop Consciousness?, Part 3: Self-Awareness," Paradigms and Revolutions, A Philosophy of Science blog, 22 Feb. 2009. http://planetparadigm.wordpress .com/2009/02/22/when-do-humans-develop-consciousness-part-3-self-awareness/.

3. Richard Woolfson, "Your Baby's Self-awareness," Made for Moms website, 29 May 2009. http://www.madeformums.com/baby-development/your-babys-self-awareness/769 .html#ixzz0lrSb3BmP.

4. R. E. Allan, *Socrates and Legal Obligation* (Minneapolis, Minn.: University of Minnesota Press, Minnesota Archive Editions, 1981), p. 3.

5. Ken Robinson, "Ken Robinson says schools kill creativity." TED Talk, June 2006. Full talk available at http://www.ted.com/talks/ken_robinson_says_schools_kill_creativity.html.

6. Basil Montagu, ed., *The Works of Francis Bacon, Lord Chancellor of England*, (London: William Pickering, 1834), vol. 16, part 2, p. ccccxlvii ("Letter to Lord Arundel").

7. "Maybe Robots Dream of Electric Sheep, But Can They Do Science?" National Science Foundation press release, 2 April 2009. http://www.nsf.gov/news/news_summ.jsp?cntn_id=114495.

8. Alan J. Perlis, "Epigrams on Programming" *SIGPLAN Notices* 17(9):7–13. http://www-pu .informatik.uni-tuebingen.de/users/klaeren/epigrams.html.

9. John Seabrooke, "It Came from Hollywood," *The New Yorker*, 1 Dec. 2003, p. 54. http:// archives.newyorker.com/?i=2003-12-01#folio=054.

10. Nick Bostrom, "Ethical Principles in the Creation of Artificial Minds," web posting 2001; rev. 2005. http://www.nickbostrom.com/ethics/aiethics.html.

11. Darren Quick, "SmartHand Brings Movement and Touch to Amputees," *Gizmag* website, 5 Nov. 2009. http://www.gizmag.com/smarthand-prosthetic-limb-touch/13277/.

12. Marvin L. Minsky, "Will Robots Inherit the Earth?" *Scientific American* online, Oct. 1994. http://web.media.mit.edu/~minsky/papers/sciam.inherit.html.

13. Joseph Needham, *Science and Civilisation in China*, vol. 2, *History of Scientific Thought* (Cambridge, UK: Cambridge University Press, 1991), p. 43.

14. "Instruction on Respect for Human Life in Its Origin and on the Dignity of Procreation," Congregation for the Doctrine of the Faith, The Vatican, 1987. http://www.vatican.va/roman_

curia/congregations/cfaith/documents/rc_con_cfaith_doc_19870222_respect-for-human-life_en.html.

Chapter 6. "Invisibly Small and Magical"

1. Stewart Brand, foreword, in K. Eric Drexler, Chris Peterson, and Gayle Pergamit, *Unbounding the Future: The Nanotechnology Revolution* (New York: William Morrow, 1991), p.5

2. Ashley Shew, "Nanotech's History: An Interesting, Interdisciplinary, Ideological Split," presented to the International Congress of Nanotechnology, 2006. http://bcs.solano.edu/ewylie/nanotech/ICNT%202006Paper.doc.

3. K. Eric Drexler, "Chapter 1: Engine of Creation: The Coming Era of Nanotechnology," online version. http://e-drexler.com/d/06/00/EOC/EOC_Chapter_1.html.

4. Pulickel M. Ajayan and Otto Z. Zhou, "Applications of Carbon Nanotubes," *Topics of Applied Physics* 80(2001):391–425. http://www.owlnet.rice.edu/~smm5/book_springer.pdf.

5. K. Eric Drexler and Richard E. Smalley, "Nanotechnology," *Chemical and Engineering News* 81(48):37–42. http://pubs.acs.org/cen/coverstory/8148/8148counterpoint.html.

6. Ashley Shew, "Nanotech's History.".

7. Ed Regis, "The Incredible Shrinking Man," *Wired*, Oct. 2004. http://www.wired.com/wired/archive/12.10/drexler.html.

8. Richard P. Feynman, "Plenty of Room at the Bottom," *Engineering and Science* (February 1960), p. 22. http://calteches.library.caltech.edu/47/3/ES.23.5.1960.0.pdf.

9. D. M. Eigler and E. K. Schweizer, "Positioning Single Atoms with a Scanning Tunnelling Microscope," *Nature* 344 (5 April 1990):524–26. http://www.nature.com/nature/journal/v344/n6266/abs/344524a0.html.

10. K. Eric Drexler, "Letter from the Author: Engine of Creation 2.0," WOWIO Books, p. 15. http://www.wowio.com/users/product.asp?BookId=503.

11. José López, "Bridging the Gaps: Science Fiction in Nanotechnology," *HYLE — International Journal for Philosophy of Chemistry*, vol. 10, no. 2 (2004), pp. 129–152. http://www.hyle.org/journal/issues/10-2/lopez.htm#ad.

12. Eric Bland, "Gold Nanospheres Sear Cancer Cells to Death," *Discovery News*, 1 April 2009. http://dsc.discovery.com/news/2009/04/01/gold-nanospheres.html.

13. Eric Hoek, "Thin Film Nanocomposites: A New Concept for Reverse Osmosis," California NanoSystems Institute website. http://www.cnsi.ucla.edu/arr/paper?paper_id=184057.

14. Munyaradzi Makani, "Nano 'tea bag' purifies water," SciDev.net, 12 Aug. 2010. http://www.scidev.net/en/news/nano-tea-bag-purifies-water.html.

15. Jaesang Lee, Shaily Mahendra and Pedro J. J. Alvarez, "Nanomaterials in the Construction Industry: A Review of Their Applications and Environmental Health and Safety Considerations," *ACS Nano*, 2010, 4 (7), pp. 3580–3590, 12 July 2010. http://pubs.acs.org/stoken/presspac/presspac/full/10.1021/nn100866w.

16. "Self-Healing House in Greece Will Dare to Defy Nature." University of Leeds Press release. http://www.issb-project.com/events/leeds.pdf.

17. The Project on Emerging Nanotechnologies—Inventories. http://www.nanotechproject.org/inventories.

18. Duncan Graham-Rowe, "'DNA Origami' Creates Map of the Americas," *New Scientist*, 15 March 2006. http://www.newscientist.com/article/dn8853-dna-origami-creates-map-of-the-americas.html.

19. Ebbe Andersen et al., "Self-assembly of a Nanoscale DNA Box with a Controllable Lid," *Nature* 459 (7 May 2009):73–76. http://www.nature.com/nature/journal/v459/n7243/full/nature07971.html.

20. Hareem Maune et al., "Self-assembly of Carbon Nanotubes into Two-dimensional Geometries Using DNA Origami Templates," *Nature Nanotechnology* 5(2010):61–66; published online, 8 Nov. 2009. http://www.nature.com/nnano/journal/v5/n1/abs/nnano.2009.311.html.

21. Richard Jones, "Even if Drexler Is Wrong, Nanotechnology Will Have Far-reaching Impacts," *SoftMachines*, 9 Nov. 2004. http://www.softmachines.org/wordpress/?p=42.

22. K. Eric Drexler, "Chapter 4: Engine of Creation: The Coming Era of Nanotechnology," online version. http://e-drexler.com/d/06/00/EOC/EOC_Chapter_4.html.

23. "Drexler and Smalley make the case for and against 'molecular assemblers,'" *Chemical and Engineering News*, vol. 81, no. 48, 1 Dec. 2003. http://pubs.acs.org/cen/coverstory/8148/8148counterpoint.html.

24. Jade Boyd, "Be a Scientist and Save the World, Says Smalley," *Rice University Press and News*, 6 Feb. 2003. http://www.media.rice.edu/media/NewsBot.asp?MODE=VIEW&ID=2039&SnID=2.

25. "Drexler and Smalley make the case for and against 'molecular assemblers'," *Chemical and Engineering News*, vol. 81, no. 48, 1 Dec. 2003. http://pubs.acs.org/cen/coverstory/8148/8148counterpoint.html.

26. Congressional hearings on nanotechnology, 1999. Prepared Written Statement and Supplemental Material of R. E. Smalley, Rice University, June 22, 1999. http://www.sc.doe.gov/bes/House/smalley.htm.

27. Ed Regis, "The Incredible Shrinking Man," *Wired*, Oct. 2004 http://www.wired.com/wired/archive/12.10/drexler.html.

28. Stewart Brand, *Whole Earth Discipline: An Eco-Pragmatist's Manifesto* (New York: Viking, 2009), "Ch.1: Scale, Scope, Stakes, Speed," p. 2.

29. Food and Agriculture Organization; International Fund for Agricultural Development; and World Food Programme of the United Nations, "Reducing Poverty and Hunger: The Critical Role of Financing for Food, Agriculture and Rural Development," report prepared for International Conference on Financing for Development, Monterrey, Mexico, 18–22 March 2002. http://www.fao.org/docrep/003/y6265e/y6265e03.htm.

30. *The State of Food Insecurity in the World 2009* (Rome: UN Food and Agriculture Organization). ftp://ftp.fao.org/docrep/fao/012/i0876e/i0876e.pdf.

31. Aaron Saenz, "Watch the Watchmen with Anti-Spy Camera Technologies," Singularity Hub website, 8 Jan. 2010. http://singularityhub.com/2010/01/08/watch-the-watchmen-with-anti-spy-camera-technologies/.

32. *Nanoscience and Nanotechnologies: Opportunities and Uncertainties* (London: The Royal Society & The Royal Academy of Engineering, July 2004), p. x. http://royalsociety.org/WorkArea/DownloadAsset.aspx?id=4294970333.

33. "Response to Council for Science and Technology's Review of Government's Progress on Nanotechnology," The Royal Society website, 24 Oct. 2006. http://royalsociety.org/Report_WF.aspx?pageid=8241.

34. Richard A. L. Jones, "Even if Drexler Is Wrong, Nanotechnology Will Have Far-reaching Impacts," *SoftMachines*, 9 Nov. 2004. http://www.softmachines.org/wordpress/?p=42.

35. Richard A. L. Jones, "Rupturing the Nanotech Rapture," *IEEE Spectrum* website, June 2008. http://spectrum.ieee.org/semiconductors/nanotechnology/rupturing-the-nanotech-rapture/2.

Chapter 7. "The Biggest, Baddest Kid on the Block"

1. "The Early Days of Pan American World Airways," Pan Am website. http://www.panamair .org/History/earlydays.htm.

2. Wilson da Silva, "Children of Apollo," *Cosmos*, Dec. 2006. http://www.cosmosmagazine .com/features/print/1163/children-apollo?page=0%2C1.

3. "Our History," Space Frontier Foundation website. http://spacefrontier.org/who-we-are/ our-history/.

4. "Frequently Asked Questions: How much does it cost to launch a Space Shuttle?" NASA: Kennedy Space Center website. http://www.nasa.gov/centers/kennedy/about/information/ shuttle_faq.html.

5. "Trek Star's Space Travel Unease," BBC News, 3 Feb. 2004. http://news.bbc.co.uk/1/hi/ entertainment/3455463.stm.

6. Dr. David Whitehouse, "Gold Rush in Space?" BBC News, 22 July 1999. http://news.bbc .co.uk/1/hi/sci/tech/401227.stm.

7. Interview with Mike Collins, in David Sington, *In the Shadow of the Moon* (Velocity/Thinkfilm, 2007).

8. *Newsweek*, 23 Dec. 1968.

9. Andrew Chaikin, "Meeting of the Minds: Buzz Aldrin Visits Arthur C. Clarke," *Space Illustrated*, 27 Feb. 2001. http://www.space.com/peopleinterviews/aldrin_clarke_010227.html.

10. Denise Chow, "Hundreds of Possible Alien Planets Discovered by NASA Spacecraft," Space .com website, 18 June 2010. http://www.space.com/scienceastronomy/kepler-spacecraft-finds-hundreds-possible-planets-100618.html.

11. Airbus Aircraft, Range of 2008 list prices document, 22 April 2008. http://www.airbus.com/ store/mm_repository/pdf/att00011726/media_object_file_ListPrices2008.pdf.

12. Spencer Reiss, "Rocket Man," *Wired*, Jan. 2005. http://www.wired.com/wired/archive/13.01/ branson.html.

13. Jeff Greason, address at International Space Development Conference, Chicago, May 27–31, 2010.

14. "SpaceX Achieves Orbital Bull's-eye with Inaugural Flight of Falcon 9 Rocket," SpaceX press release, 7 June 2010. http://www.spacex.com/press.php?page=20100607.

15. "Meet the teams," Google Lunar X prize website. http://www.googlelunarxprize.org/lunar/ teams.

16. "A Comparison of Risk: Accidental Deaths, United States 1999–2003," 1 Jan. 2004. U.S. Department of Transportation, Pipeline and Hazardous Materials Safety Administration website. http://www.phmsa.dot.gov/portal/site/PHMSA/menuitem.ebdc7a8a7e39f2e55cf2031 050248a0c/?vgnextoid=8524adbb3c60d110VgnVCM1000009ed07898RCRD&vgnextchann el=4f347fd9b896b110VgnVCM1000009ed07898RCRD&vgnextfmt=print.

17. James Michael Snead, "Achieving Near-term, Aircraft-like Reusable Space Access," Air Force Research Laboratory, Wright-Patterson Air Force Base, n.d. http://mikesnead.net/resources/ spacefaring/tech_paper_achieving_near-term_aircraft-like_reusable_space_access.pdf.

18. Larry Dignan, "Virgin Galactic Unveils SpaceShipTwo: Plans Open Architecture Spaceship," ZDNet: News and Blogs, 23 Jan. 2008. http://www.zdnet.com/blog/btl/virgin-galactic-unveils-spaceshiptwo-plans-open-architecture-spaceship/7678.

19. Dick Rutan website: http://www.dickrutan.com/voyager.html.

Chapter 8. "A Constant and Complete Intercourse"

1. Malcolm Bacchus, "The Telegraph on Telegraph Hill," Telegraph Hill Society, 2010. http://thehill.org.uk/society/Telegraph.htm.

2. Gerard J. Holzmann and Bjoern Pehrson, "The First Data Networks," *Scientific American*, Jan. 1994.

3. The Institute of Telecom Resources EMeIAAA. http://www.itre.org/Quotes.htm.

4. Steven Pinker, "A History of Violence," Edge World Question Center website. http://www.edge.org/3rd_culture/pinker07/pinker07_index.html.

5. Steven Pinker, *The Blank Slate: The Modern Denial of Human Behavior* (New York: Penguin, 2002), p. 57.

6. "Annex Table 2 - Deaths by cause, sex and mortality stratum in WHO regions, estimates for 2002," *World Health Organization 2004 Report* (Geneva: UN World Health Organization). http://www.who.int/whr/2004/annex/topic/en/annex_2_en.pdf.

7. Human Security Brief 2006, Human Security Center, University of British Columbia. http://www.humansecuritybrief.info/2006/contents/finalversion.pdf.

8. Ziad Obermeyer, Christopher J. L. Murray and Emmanuela Gakidou, "Fifty Years of Violent War Deaths from Vietnam to Bosnia: Analysis of Data from the World Health Survey Program," *British Medical Journal* 336(28 June 2008):1482–86. http://www.bmj.com/cgi/content/full/bmj.a137.

9. "About Us," Oxford Research Group website. http://www.oxfordresearchgroup.org.uk/about_us.

10. John Sloboda et al., "Identifying and Recording Every Casualty of Armed Conflict: A Discussion Document," © Oxford Research Group 2007. http://www.oxfordresearchgroup.org.uk/sites/default/files/CTC Consultation.pdf.

11. Robert Wright, "Optimism," TED talk, 2006. Full talk available at http://www.ted.com/talks/robert_wright_on_optimism.html.

12. "Net Turns 30," BBC News, 20 Oct. 1999. http://news.bbc.co.uk/1/hi/sci/tech/480102.stm.

13. John Chu, "In the Internet age, dance evolves . . . ," TED talk, February 2010. Full talk available at http://www.ted.com/talks/the_lxd_in_the_internet_age_dance_evolves.html.

14. Seth Godin, "The Tribes We Lead," TED talk, Feb. 2009. Full talk available at http://www.ted.com/talks/seth_godin_on_the_tribes_we_lead.html.

15. Author interview with John Seely Brown, 16 April 2010.

16. Gabriel Weimann, "www.terror.net: How Modern Terrorism Uses the Internet," United States Institute of Peace, March 2004. http://www.usip.org/resources/wwwterrornet-how-modern-terrorism-uses-internet.

17. "YouTube Shutdown in Pakistan," Pakistan Telecommunication Authority press release, 27 May 2010. http://www.pta.gov.pk/index.php?option=com_content&task=view&id=1391&Itemid=301.

18. "Internet Enemies, Countries under Surveillance, Australia," Reporters Without Borders website. http://en.rsf.org/surveillance-australia,36674.html.

19. Asher Moses, "Queensland Dentist Furious over Blacklist," *Brisbane Times*, 19 March 2009. http://www.brisbanetimes.com.au/articles/2009/03/19/1237054974422.html.

20. "Australian Government Secret ACMA Internet Censorship Blacklist," WikiLeaks, 6 August 2008. http://wikileaks.org/wiki/Australian_government_secret_ACMA_internet_censorship_blacklist,_6_Aug_2008.

21. "The Top 500 Sites on the Web," Alexa website, June 2010. http://www.alexa.com/topsites.

22. "Expert Report of Philip B. Stark," 8 May 2006, *ACLU* v. *Gonzales*, Civ. Action No. 98-5591 (E.D. Pa.). http://sethf.com/infothought/blog/archives/copa-censorware-stark-report.pdf.

23. WikiQuote, "Stephen Hawking." http://en.wikiquote.org/wiki/Stephen_Hawking.

24. Vint Cerf, "The Next Internet," Official Google blog, 25 Sept. 2008. http://googleblog .blogspot.com/2008/09/next-internet.html.

25. "Memorable Quotes for *Connections*," Episode 1: "The Trigger Effect," 17 Oct. 1978, Internet Movie Data Base © BBC. http://www.imdb.com/title/tt0818341/quotes.

26. W. Daniel Hillis, "The Dawn of the Entanglement," *Edge World Question Center.* http://www .edge.org/q2010/q10_1.html.

27. E. M. Forster, "The Machine Stops," Ebook at http://manybooks.net/titles/forstereother07 machine_stops.html.

28. "Pattie Maes and Pranav Mistry Demo SixthSense," TED talk, February 2009. Full talk available at http://www.ted.com/talks/pattie_maes_demos_the_sixth_sense.html.

29. "Blaise Aguera y Arcas Demos Augmented-reality Maps," TED talk, Feb. 2010. Full talk available at http://www.ted.com/talks/blaise_aguera.html.

30. "Interview with James Lovelock," *Today*, BBC Radio 4, 30 March 2010. http://news.bbc .co.uk/today/hi/today/newsid_8594000/8594274.stm.

Chapter 9. "World Leaders Still Don't Really Get It"

1. "Mauna Loa CO_2 Monthly Mean Data, Earth System Research Laboratory—Global Monitoring Division," U.S. National Oceanic & Atmospheric Administration website. ftp://ftp.cmdl .noaa.gov/ccg/co2/trends/co2_mm_mlo.txt.

2. Goddard Institute for Space Studies Surface Temperature Analysis, Jan. 2010, at http://data .giss.nasa.gov/gistemp/paper/gistemp2010_draft0601.pdf.

3. J.-M. Barnola et al., "Historical Carbon Dioxide Record from the Vostok Ice Core," Carbon Dioxide Information Analysis Center. http://cdiac.ornl.gov/trends/co2/vostok.html.

4. "Fewer Americans See Solid Evidence of Global Warming," Pew Research Center for the People & the Press, 22 Oct. 2009. http://people-press.org/report/556/global-warming.

5. Simon Jackman, "Australians, Americans and Climate Change" (Sydney, Australia: U.S. Studies Centre, Nov. 2009). http://ussc.edu.au/s/media/docs/publications/091120_climate _change_survey.pdf.

6. Owen Bowcott, "Climate change concern declines in poll," The Guardian, 23 May 2010. http:// www.guardian.co.uk/environment/2010/may/23/climate-change-interest-yougov-survey.

7. François-Marie Bréon, "How Do Aerosols Affect Cloudiness and Climate?" *Science* 313(5787):623–24. http://www.sciencemag.org/cgi/content/short/313/5787/623.

8. Professor John Latham, "Futuristic fleet of 'cloudseeders'—Viewpoint," BBC News, 15 Feb. 2007. http://news.bbc.co.uk/1/hi/programs/6354759.stm.

9. Mary Hardin and Ralph Kahn, "Aerosols and Climate Change," NASA Earth Observatory website. http://earthobservatory.nasa.gov/Features/Aerosols/aerosol2.php.

10. Alan Robock, "Mount Pinatubo as a Test of Climate Feedback Mechanisms," *Volcanism and the Earth's Atmosphere*, Geophysical Monograph 139, © 2003 American Geophysical Union. http://climate.envsci.rutgers.edu/pdf/VEAChapter1_Robocknew.pdf.

11. "NASA Finds Soot Has Impact on Global Climate," Goddard Institute for Space Studies/ NASA website, 13 May 2003. http://www.giss.nasa.gov/research/news/20030513/.

12. "Ch. 6: Reducing Black Carbon," in *State of the World 2009: Into a Warming World*, 26th ed. (Washington, D.C.: Worldwatch Institute).

13. Mike Berners-Lee, *How Bad Are Bananas? The Carbon Footprint of Everything*. (London: Profile Books, 2010).

14. Quirin Schiermeier, "The Real Holes in Climate Science," *Nature*, 463(20 Jan. 2010):284–87. http://www.nature.com/news/2010/100120/full/463284a.html.

15. "Orbiting Carbon Observatory," NASA website at http://oco.jpl.nasa.gov/science/. And also Jonathan Amos, "Failure Hits Nasa's 'CO2 Hunter,'" BBC News, 24 Feb. 2009. http://news.bbc.co.uk/1/hi/sci/tech/7907570.stm.

16. Lauren Morello, "Scientists Mull Future After Carbon Satellite Crash," *New York Times*, 25 Feb. 2009. http://www.nytimes.com/cwire/2009/02/25/25climatewire-after-carbon-observatory-crashes-scientists-a-9853.html.

17. Leo Hickman, "James Lovelock on the Value of Skeptics and Why Copenhagen Was Doomed," *Guardian*, 29 March 2010. http://www.guardian.co.uk/environment/blog/2010/mar/29/james-lovelock.

18. Commonwealth of Australia, *Parliamentary Debates*, House of Representatives Official Hansard, no. 11, 11 Aug. 2009. http://parlinfo.aph.gov.au/parlInfo/download/chamber/hansardr/2009-08-11/toc_pdf/7007-3.pdf;fileType=application%2Fpdf#search=%22chamber/hansardr/2009-08-11/0000%22.

19. *The Daily Show*, 29 Oct. 2009. http://www.thedailyshow.com/watch/thu-october-29-2009/exclusive—al-gore-extended-interview-pt—1.

20. Stewart Brand, "Scale, Scope, Stakes and Speed: Ecopragmatic Heresy & the Long-Term Future of Humanity," presented to the Institute of Contemporary Arts, London, 30 Jan. 2010.

21. Peter T. Doran and Maggie Kendall Zimmerman, "Examining the Scientific Consensus on Climate Change," *Eos* 90(3). http://tigger.uic.edu/~pdoran/012009_Doran_final.pdf.

22. "Earth Institute Profile: Wallace S. Broecker," Earth Institute, Columbia University website. http://www.earth.columbia.edu/articles/view/2246.

23. David Archer, *The Long Thaw: How Humans Are Changing the Next 100,000 Years of Earth's Climate* (Princeton, N.J.: Princeton University Press, 2008).

24. "Humidity Relative to Earth's Temperature," NASA website, 20 April 2007. http://www.nasa.gov/vision/earth/lookingatearth/warmer_humidity.html.

25. Bert Metz, "IPCC Special Report on Carbon Dioxide Capture and Storage," Intergovernmental Panel on Climate Change (Working Group III), p. 108. http://www.ipcc.ch/pdf/special-reports/srccs/srccs_chapter3.pdf.

26. "Scientists Target East Coast Rocks for CO2 Storage," Earth Institute, Columbia University, website, 4 Jan. 2010. http://www.earthinstitute.columbia.edu/articles/view/2611.

27. Richard M. Jones, "Uncertain Outlook for DOE's Energy Innovation Hubs," *FYI: The American Institute of Physics Bulletin of Science Policy News*, 20 July 2009. http://www.aip.org/fyi/2009/095.html.

28. Home page, Novomer website. http://www.novomer.com.

29. Stuart Licht et al., "A New Solar Carbon Capture Process: Solar Thermal Electrochemical Photo (STEP) Carbon Capture," supporting information for publications prepared by Department of Chemistry, George Washington University and Department of Chemical Engineering, Howard University. http://pubs.acs.org/doi/suppl/10.1021/jz100829s/suppl_

file/jz100829s_si_001.pdf. Full paper available at http://pubs.acs.org/doi/abs/10.1021/jz100829s.

30. "Alberta Researchers Claim Their Machine Removes CO2 from Air," *Calgary Herald*, 29 Sept. 2008. http://www.canada.com/topics/news/national/story.html?id=a6e92755-194c-4bf1-890b-ad52be9954ec.

31. "Nonconventional Liquid Fuels," U.S. Energy Information Administration website, 2006. http://www.eia.doe.gov/oiaf/aeo/otheranalysis/aeo_2006analysispapers/nlf.html.

32. "Slide 8: Improvements in Ethanol Efficiency," *American Ethanol in the World Market*, Renewable Fuels Association presentation at World Biofuels conference, Seville, Spain, 19 May 2010. http://www.ethanolrfa.org/page/-/World Biofuels 2010.pdf.

33. Brian L. Goodall, "Algae: The Source of Reliable, Scalable and Sustainable Liquid Transportation Fuels," presented at joint Web conference of U.S. Department of Energy's Biomass and Clean Cities programs, 12 Feb. 2009. http://www1.eere.energy.gov/biomass/pdfs/goodall_20090212.pdf.

34. Martin LaMonica, "Algae Farm in Mexico to Produce Ethanol in '09," *cnet News*, 12 June 2008. http://news.cnet.com/8301-11128_3-9966867-54.html.

35. "Grant Amounts," U.S. Department of Energy website. www.energy.gov/media/564M_Biomass_Projects.pdf.

36. "Frequently Asked Questions," Joule Unlimited website. http://www.jouleunlimited.com/faq.

37. "Frequently Asked Questions: Gasoline," figures from U.S. Energy Information Administration website. http://www.eia.doe.gov/ask/gasoline_faqs.asp.

38. "Who Are the Major Players Supplying the World Oil Market?" U.S. Energy Information Administration website. http://www.eia.doe.gov/energy_in_brief/world_oil_market.cfm.

39. "How dependent are we on foreign oil?" U.S. Energy Information Administration website. http://www.eia.doe.gov/energy_in_brief/foreign_oil_dependence.cfm.

40. "What Is a Flexible Fuel Vehicle?" U.S. Department of Energy, Alternative and Advanced Vehicles website. http://www.afdc.energy.gov/afdc/vehicles/flexible_fuel_what_is.html.

41. "GM, National Governors Association Team Up on E85: Goal Is Building Infrastructure for Next-Generation Biofuels," General Motors press release, 17 July 2008. http://archives.media.gm.com/archive/documents/domain_2/docId_47043_pr.html.

Chapter 10. "I'm a Disruptive Guy. I Like Disruption."

1. "Fast Facts," Solarbuzz website. http://www.solarbuzz.com/fastfactsindustry.htm.

2. *Popular Science Monthly,* April 1877; trans. from *Revue des Deux Mondes,* 1 May 1876.

3. Paul Collins, "The Beautiful Possibility," *Cabinet* 6 (Spring 2002). http://www.cabinetmagazine.org/issues/6/beautifulpossibility.php.

4. "Concept: The Technologies," Desertec Foundation website. http://www.desertec.org/en/concept/technologies/.

5. "Desertec: Broad Support from All Sectors of the Economy," Desertec press release, 28 April 2010. http://www.dii-eumena.com/fileadmin/Daten/press/10-04-28_Dii_Associated_Partners_EN.pdf.

6. "Electric Generation Efficiency," Topic Paper #4, *NPC Global Oil & Gas Study*, 18 July 2007. http://www.npc.org/Study_Topic_Papers/4-DTG-ElectricEfficiency.pdf.

7. "Where can I find data on electricity transmission and distribution losses? Frequently Asked Questions: Electricity," U.S. Energy Information Administration website. http://tonto.eia.doe.gov/ask/electricity_faqs.asp#electric_rates2.

8. "Advanced Technologies and Energy Efficiency," U.S. Department of Energy. http://www.fueleconomy.gov/feg/atv.shtml.

9. Tsigue Shiferaw, "Sun Energy Empowers Ethiopian Village," BBC News, 16 July 2009. http://news.bbc.co.uk/1/hi/world/africa/8150391.stm.

10. Mark Jaffe, "Xcel Rate-hike Plan Slammed at PUC," *Denver Post*, 6 Aug. 2009. http://www.denverpost.com/breakingnews/ci_13001159.

11. Matthew Philips, "Taking a Dim View of Solar Energy," *Newsweek*, 25 Aug. 2009. http://www.newsweek.com/content/newsweek/2009/08/24/taking-a-dim-view-of-solar-energy.html.

12. "Improving Lives: World Bank Group Progress on Renewable Energy and Energy Efficiency, Fiscal Year 2006," report by the World Bank, Dec. 2006. http://siteresources.worldbank.org/EXTENERGY/Resources/336805-1157034157861/Improving_Lives_Low_Res.pdf.

13. Dan Boniface, "Colo. Town Energized to Go Off Electric Grid, *9 News*, May 2010. http://www.9news.com/money/article.aspx?storyid=137546&catid=531.

14. "Nanotech Batteries for a New Energy Future," *Science Daily*, 22 March 2009. http://www.sciencedaily.com/releases/2009/03/090320173859.htm.

15. "Springs Built from Nanotubes Could Provide Big Power Storage Potential," Massachusetts Institute of Technology press release, 21 Sep. 2009. http://web.mit.edu/newsoffice/2009/super-springs-0921.html.

16. "International Energy Outlook 2010—Highlights," U.S. Energy Information Administration website. http://www.eia.doe.gov/oiaf/ieo/highlights.html.

17. "Nanosolar Utility Panel Data Sheet, March 2010," Nanosolar website. http://www.nanosolar.com/sites/default/files/Nanosolar_Utility_Panel_Data_Sheet_r2.0.pdf.

18. Tom Cheyney, "Printed PV: Nanosolar Unveils 640MW Utility-scale Panel Fab, High-efficiency CIGS Cell Production," PV-Tech website, 9 Sept. 2009. http://www.pv-tech.org/news/_a/printed_pv_nanosolar_unveils_640mw_utility-scale_panel_fab_high-efficiency_/.

19. "Nanosolar Completes Panel Factory, Commences Serial Production," Nanosolar press release, 9 Sept. 2009. http://www.nanosolar.com/company/blog/nanosolar-completes-panel-factory-commences-serial-production.

20. Todd Woody, "$4.1 Billion in Orders for Thin-Film Solar," *New York Times,* "Green" blog, 9 Sept. 2009. http://green.blogs.nytimes.com/2009/09/09/41-billion-in-orders-for-thin-film-solar/.

21. Dan Lewis, "Solar Grid Parity," *Engineering & Technology* 9, 19 May 2009. http://kn.theiet.org/magazine/issues/0909/solar-grid-parity-0909.cfm.

22. "Fast Solar Energy Facts," Solarbuzz website. http://www.solarbuzz.com/fastfactsindustry.htm.

23. "Solarbuzz Reports World Solar Photovoltaic Market Grew to 7.3 Gigawatts in 2009," Solarbuzz website. http://www.solarbuzz.com/Marketbuzz2010-intro.htm.

Chapter 11. "The Black Phantom"

1. Gaia Vince, "One Last Chance to Save Mankind: Interview with James Lovelock," *New Scientist*, 23 Jan. 2009. http://www.newscientist.com/article/mg20126921.500-one-last-chance-to-save-mankind.html.

2. Ibid.

3. "Soil Carbon Sequestration Fundamentals," Ohio State University fact sheet, n.d. http://ohioline.osu.edu/aex-fact/0510.html.

4. Kris De Decker, "Wood Gas Vehicles: Firewood in the Fuel Tank," *Low-tech Magazine*, 18 Jan. 2010. http://www.lowtechmagazine.com/2010/01/wood-gas-cars.html.

5. J. Amonette, J. Lehmann and S. Joseph, "Terrestrial Carbon Sequestration with Biochar: A Preliminary Assessment of Its Global Potential," presented at American Geophysical Union, Fall 2007. http://adsabs.harvard.edu/abs/2007AGUFM.U42A.06A.

6. Vince, "One Last Chance to Save Mankind."

7. "Aquaflow Wild Algae Converted to Key Jet Fuel Component—World-first Breakthrough," Aquaflow press release, 4 Dec. 2009. http://www.aquaflowgroup.com/wp-content/uploads/2010/04/media_release_14_12_2008.pdf.

8. "Aquaflow Secures Role in Second U.S. Department of Energy-funded Algal Project," Aquaflow press release, 10 March 2010. http://www.aquaflowgroup.com/wp-content/uploads/2010/04/media_release_10_5_2010.pdf.

9. Bill Dunn, "The New Victorians, *The Sunday Times*, 5 Dec. 2009. http://www.timesonline.co.uk/tol/news/science/article6937504.ece.

10. Paul Blackwell, Glen Riethmuller and Mike Collins, *Biochar for Environmental Management: Science and Technology* (London: Earthscan Publications, 2009), "Ch. 12: Biochar Application to Soil."

11. A. Gathorne-Hardy, J. Knight and J. Woods, "Biochar as a Soil Amendment Positively Interacts with Nitrogen Fertiliser to Improve Barley Yields in the UK," IOP Conference Series: *Earth and Environmental Science* 6, Session 37, 2009. http://iopscience.iop.org/1755-1315/6/37/372052/pdf/1755-1315_6_37_372052.pdf.

12. "Biochar Soil Management: Basic Concepts," Cornell University, Department of Crop and Soil Sciences website: http://www.css.cornell.edu/faculty/lehmann/research/biochar/biocharmain.html.

13. George Monbiot, "Woodchips with Everything," *Guardian*, 24 March 2009. http://www.guardian.co.uk/environment/2009/mar/24/george-monbiot-climate-change-biochar.

14. Kurt Kleiner, "The Bright Prospect of Biochar," *Nature* Reports: Climate Change, 21 May 2009. http://www.nature.com/climate/2009/0906/full/climate.2009.48.html.

15. "Shreddies Flatulence Filtering Underwear," Shreddies website. http://www.myshreddies.com/flatulence-underwear/?a=cm.

Chapter 12. "A Little Bit of a Big Amount"

1. Allan Savory, with Jody Butterfield, *Holistic Management* (Washington, D.C.: Island Press, 2nd ed., 1999), "Ch. 5: The Predator-Prey Connection." The original passage, which I've paraphrased, reads: "Over the next few years I spent thousands of hours tracking people over all sorts of country, day after day. This discipline greatly sharpened my observational skills and also taught me much about the land, as although I was hunting humans, my thoughts were constantly on the state of the lands over which we were fighting. I doubt many scientists ever had such an opportunity for learning. I tracked people over game areas, tribal areas, commercial farms and ranches, and over all different soil and vegetation types in all rainfalls. Often I covered many different areas in a single day as I flew by helicopter from one trouble spot to the next. Everywhere I had to inspect plants and soils for the faintest sign of disturbance by people trying to leave no hint of their passage."

2. Saffron Hosden and Jessica Mahar, "Floodwater fail to stop drought's creep across state," *Sydney Morning Herald*, 12 Jan. 2010. http://www.smh.com.au/environment/floodwaters-fail-to-stop-droughts-creep-across-state-20100111-m2s6.html.

3. N. Burgess, "Agricultural Debt from 1994 to 2009," sourced from Westpac Economics & Reserve Bank of Australia, 2010.

4. "Assistance Guide: Drought Support Workers," New South Wales Government pamphlet, August 2010. http://www.dpi.nsw.gov.au/agriculture/emergency/drought/assistance/support/nsw-drought-assistance-guide.

5. Sandra Kanck, "Gillard on Population: An Outbreak of Commonsense?" Sustainable Population Australia media release, 28 June 2010. http://www.population.org.au/index.php/media/media-releases/media-releases-2010/552-gillard-on-population-an-outbreak-of-commonsense.

6. "The Manchester Report," *The Guardian*, July 2009, p. 13. http://www.guardian.co.uk/environment/manchester-report.

7. Transcript, "Episode 3," *The Decade Old Headline*, Australia Broadcasting Corporation Television MediaWatch, 2007. http://www.abc.net.au/mediawatch/transcripts/s1869891.htm.

8. K. Carrington, A. McIntosh and J. Scott, "Date Report No. 1: Secondary Data Analysis Regarding Self-harm Including Suicide," *Australian Research Council Discovery Project Safeguarding Rural Australia: Addressing Masculinity and Violence in Rural Settings*, Nov. 2009. http://www.aph.gov.au/Senate/committee/clac_ctte/suicide/submissions/sub62a.pdf.

9. "New South Wales State and Regional Indicators," Australian Bureau of Statistics, Dec. 2009. http://www.abs.gov.au/ausstats/abs@.nsf/Products/1338.1~Dec+2009~Main+Features~Environment?OpenDocument.

10. "About the Stubbyglove," Stubbyglove website. http://www.freewebstore.org/Stubbyglove Enterprises/storepage225698.aspx.

11. William Ruddiman, *Plows, Plagues, and Petroleum: How Humans Took Control of Climate* (Princeton, N.J.: Princeton University Press, 2005).

12. Henning Steinfeld et al., "Livestock's Long Shadow: Environmental Issues and Options," *Livestock, Development and Environment Initiative* (Rome: UN Food and Agricultural Organization, 2006). ftp://ftp.fao.org/docrep/fao/010/a0701e/a0701e00.pdf.

13. Ibid., p. 95.

14. "CO2 Emissions from Fuel Combustion: Highlights" (Paris, France: International Energy Agency, 2009). http://www.iea.org/co2highlights/CO2highlights.pdf.

15. Ann Adams, "Carbon Neutral Ranching: Blackstone Ranch," *Land and Livestock Magazine* (March/April 2010):15–16.

16. Eleonora Panunzi, "Are Grasslands Under Threat? Brief Analysis of FAO Statistical Data on Pasture and Fodder Crops" (Rome: UN Food and Agriculture Organization, July 2008). http://www.fao.org/ag/agp/agpc/doc/grass_stats/grass-stats.htm.

17. Tiago Domingos, "Carbon Sequestration in Sown Biodiverse Pasture," *Project Terraprima— Portuguese Carbon Fund*, presented at UN Climate Change Conference, Copenhagen, 10 Dec. 2009.

18. Cost of beef production, 2009 figure from Holmes and Sackett. http://www.hs-a.com.au/index.htm.

19. Peter Donovan, "The Calculation: Soil Organic Matter Needed to Bring Down Atmospheric Carbon," Soil Carbon Coalition website: http://www.soilcarboncoalition.org/calculation.

Chapter 13. "The Best Job in the World."

1. Wendy Zukerman, "Shape-shifting Islands Defy Sea-level Rise," *New Scientist*, 2 June 2010. http://www.newscientist.com/article/mg20627633.700-shapeshifting-islands-defy-sealevel-rise.html.

2. Marc Kaufman, "Escalating Ice Loss Found in Antarctica: Sheets Melting in an Area Once Thought to Be Unaffected by Global Warming," *Washington Post*, 14 Jan. 2008. http://www.washingtonpost.com/wp-dyn/content/article/2008/01/13/AR2008011302753_pf.html.

3. J. Bamber et al., "Reassessment of the Potential Sea-Level Rise from a Collapse of the West Antarctic Ice Sheet," *Science*, 15 May 2009. http//www.sciencemag.org/cgi/content/abstract/324/5929/901.

4. Rob Young and Orrin Pilkey, "Get Ready for Seven-foot Sea Level Rise as Climate Change Melts Ice Sheets," *Guardian Yale Environment 360*, 15 Jan. 2010. http://www.guardian.co.uk/environment/2010/jan/15/sea-level-climate-change.

5. Helmut Schuhmacher et al., "The Aftermath of Coral Bleaching on a Maldivian Reef: A Quantitative Study," *Facies* 51(1–4):80–92. http://www.springerlink.com/content/k6788j1h36qu6702/.

6. Alan Wheatley, "Maldives Nurses Its Coral Reefs Back to Life," Global Coral Reef Alliance website, 2 May 2004. http://www.globalcoral.org/Maldives Nurses Its Coral Reefs Back to Life.htm.

7. Wade Norris, "A Human Rights Petition to President Barack Obama," *Huffington Post*, 28 Oct. 2009. http://www.huffingtonpost.com/wade-norris/a-human-rights-petition-t_b_336074.html.

8. President Mohamed Nasheed of the Republic of the Maldives, "Address" presented at UN Summit on Climate Change. http://www.un.org/News/Press/docs/2009/envdev1069.doc.htm.

9. President Mohamed Nasheed of the Republic of the Maldives, "Is Gandhi Relevant Today?" keynote address to mark International Non-Violence Day, 19 Oct. 2009. http://www.presidencymaldives.gov.mv/4/?ref=1,6,2256.

10. President Mohamed Nasheed of the Republic of the Maldives, address to the UN General Assembly, 24 Sept. 2009. http://www.presidencymaldives.gov.mv/4/?ref=1,6,2052.

11. "Maldives: Wave of Arrests and Uninvestigated Deaths Suggest Human Rights Abuse Growing," Amnesty International website, 23 Sept. 2003. http://www.amnesty.org.uk/news_details.asp?NewsID=14901.

12. President Mohamed Nasheed of the Republic of the Maldives, address to U.K. Conservative Party Conference, 8 Oct. 2009. http://www.livestream.com/conservatives/video?clipId=flv_ede41f98-bf4d-4b33-b81e-c70443aec3c0.

13. Christine Toomey, "The Maldives: Trouble in Paradise," *The Times* online, 1 Feb. 2009. http://www.timesonline.co.uk/tol/news/environment/article5604464.ece.

14. Rahul Jacob, "Afternoon Tea with the Financial Times: Mohamed Nasheed," *Financial Times*, 27 Nov. 2009. http://www.ft.com/cms/s/2/35f6ff72-dae2-11de-933d-00144feabdc0.html.

15. Peter Foster, "Maldives Dictator Tries a Touch of New Labour Spin," *Telegraph*, 17 June 2006. http://www.telegraph.co.uk/news/worldnews/africaandindianocean/maldives/1521549/Maldives-dictator-tries-a-touch-of-New-Labour-spin.html.

16. CIA World Factbook, "Southeast Asia: Maldives. https://www.cia.gov/library/publications/the-world-factbook/geos/mv.html.

17. Judith Evans, "Maldives Ruler Voted Out of Office after 30 years," *Reuters*, 29 Oct. 2008. http://www.alertnet.org/thenews/newsdesk/COL392374.htm.

18. President Mohamed Nasheed of the Republic of the Maldives, address to UN Summit on Climate Change, 22 Sept. 2009. http://www.presidencymaldives.gov.mv/4/?ref=1,6,1978.

19. http://www.populstat.info/Asia/maldivec.htm.

20. Worldmapper project, University of Sheffield, UK. http://www.worldmapper.org/.

21. "World Urbanization Prospects: The 2007 Revision" (New York: United Nations Population Division, 2008). http://www.un.org/esa/population/publications/wup2007/2007WUP _Highlights_web.pdf.

22. Thomas Malthus, "An Essay on the Principle of Population as It Affects the Future Improvement of Society" (London: J. Johnson, 1798). http://www.econlib.org/library/Malthus/ malPop1.html.

23. "How the World Is Fed," Ch. 2 of *Agriculture, Food and Water* (Rome: UN Food and Agriculture Organization, 2003). http://www.fao.org/docrep/006/y4683e/y4683e06.htm.

24. Hans Rosling, "What Stops Population Growth?" a presentation on Gapminder website: http://www.gapminder.org/videos/what-stops-population-growth/.

25. *World Population Prospects, The 2002 Revision* (New York: UN Publications, United Nations Population Division, Feb. 2003). http://www.un.org/esa/population/publications/wpp2002/ WPP2002-HIGHLIGHTSrev1.PDF.

26. "Maldives: Statistics," Demographic Indicators, Total fertility rate, 2008. UNICEF website. http://www.unicef.org/infobycountry/maldives_maldives_statistics.html.

27. Total fertility rate (children per woman). United Nations database website. http://data .un.org.

28. George Martine, "The State of the World's Population: Unleashing the Potential of Urban Growth," presented at the Woodrow Wilson International Center for Scholars, 27 June 2007. http:// www.wilsoncenter.org/index.cfm?topic_id=116811&fuseaction=topics.event_summary& event_id=239720.

29. *State of the World's Forests 2009* (Rome: UN Food and Agriculture Organization, 2010). http:// www.fao.org/docrep/011/i0350e/i0350e00.HTM.

30. Elisabeth Rosenthal, "New Jungles Prompt a Debate on Rain Forests," *New York Times*, 29 Jan. 2009. http://www.nytimes.com/2009/01/30/science/earth/30forest.html.

31. "World leaders accept Copenhagen Accord," Website of The Maldives Mission to the European Union, 20 Dec. 2009. http://www.maldivesmission.eu/index.php?option= com_content&view=article&id=238:20th-december-2009&catid=33:december&Itemid=30.

Chapter 14. "Making a Road Where There Isn't One"

1. Ray Kurzweil, *The Singularity Is Near: When Humans Transcend Biology* (New York: Viking, 2005), "Ch. 1: The Six Epochs."

2. Stewart Brand, *Whole Earth Discipline* (New York: Viking, 2009), "Ch. 1: Scale, Scope, Stakes, Speed," p. 2.

3. Matt Ridley, *The Rational Optimist* (New York: HarperCollins, 2010), "Ch. 1: When Ideas Have Sex."

4. Greg Ross, "An Interview with Douglas R. Hofstadter,"*American Scientist*, n.d. http://www .americanscientist.org/bookshelf/pub/douglas-r-hofstadter.

5. James Shreeve, "Craig Venter's Epic Voyage to Redefine the Origin of the Species," *Wired* 12(8). http://www.wired.com/wired/archive/12.08/venter.html.

6. Juan Enriquez, *As the Future Catches You* (New York: Three Rivers Press, 2001), "Ch. 1: Mixing Apples, Oranges, and Floppy Disks . . . ", p. 12.

Chapter 15. "Future Shock"

1. John Seely Brown, "The Entrepreneur and the Cloud: Silicon Valley Rejuventated," Stanford University online video presentation, 24 Sept. 2009. http://sprie.stanford.edu/events/recording/5790/1/306.

2. John Hagel III, John Seely Brown and Lang Davison, *The Power of Pull: How Small Moves, Smartly Made, Can Set Big Things in Motion* (New York: Basic Books, 2010), "Ch. 1: The Diminishing Power of Push."

3. Ken Robinson, "Schools Kill Creativity," TED talk, Feb. 2006. Full talk available at http://www.ted.com/talks/lang/eng/ken_robinson_says_schools_kill_creativity.html.

4. Seth Godin, "The Tribes We Lead," TED talk, Feb. 2009. Full talk available at http://www.ted.com/talks/seth_godin_on_the_tribes_we_lead.html.

5. Chris Anderson, "Chris Anderson Shares His Vision for TED," TED talk, Feb. 2002. Full talk available at http://www.ted.com/talks/chris_anderson_shares_his_vision_for_ted.html.

Chapter 16. Hello, and Thanks for All the Fish.

1. Letter to the Editor re Mössner, Müller-Vahl, Döring and Stuhrmann, "Role of the Novel Tryptophan Hydroxylase-2 Gene in Tourette syndrome," *Molecular Psychiatry* 12(July 2007):617–619. http://www.nature.com/mp/journal/v12/n7/full/4002004a.html.

2. M. D. Brown et al. "Promotion of Prostatic Metastatic Migration Toward Human Bone Marrow Stoma by Omega 6 and Its Inhibition by Omega 3 PUFAs," *British Journal of Cancer* 94(6):842–53. http://www.ncbi.nlm.nih.gov/pubmed/16523199.

3. *Diarrhea: Why Children Are Still Dying and What Can Be Done* (Geneva: UNICEF/World Health Organization, 2009). http://www.who.int/child_adolescent_health/documents/9789241598415/en/index.html.

4. Malaria fact sheet, World Health Organization. http://www.who.int/mediacenter/factsheets/fs094/en/index.html.

5. "Rwanda: How the Genocide Happened," BBC News, 18 Dec. 2008. http://news.bbc.co.uk/1/hi/world/africa/1288230.stm.

6. Hitha Prabhakar, "World's Most Extravagant Handbags," *Forbes*, 3 Aug. 2007. http://www.forbes.com/2007/03/26/handbags-extravagant-expensive-forbeslife-cx_hp_0327handbags.html.

7. Matt Ridley, *The Rational Optimist: How Prosperity Evolves* (New York: HarperCollins, 2010), "Ch. 9: Turning Points: Pessimism After 1900."

8. Romain Rolland, "Review of R. Lefebvre's *The Sacrifice of Abraham*," *L'Humanité*, 19 March 1920.

INDEX